17th International School-Conference for Young Scientists and Specialists "New Materials: Advanced Technologies"

IOP Conference Series: Materials Science and Engineering
Volume 1005

Moscow, Russia
5 - 8 November 2019

ISBN: 978-1-7138-3317-8
ISSN: 1757-8981

Printed from e-media with permission by:

Curran Associates, Inc.
57 Morehouse Lane
Red Hook, NY 12571

Some format issues inherent in the e-media version may also appear in this print version.

Institute of Physics
Dirac House, Temple Back
Bristol BS1 6BE UK

Phone: 44 1 17 929 7481
Fax: 44 1 17 920 0979

techtracking@iop.org

Additional copies of this publication are available from:

Curran Associates, Inc.
57 Morehouse Lane
Red Hook, NY 12571 USA
Phone: 845-758-0400
Fax: 845-758-2633
Email: curran@proceedings.com
Web: www.proceedings.com

TABLE OF CONTENTS

IOP Conf. Series: Materials Science and Engineering 1005 (2020) 011001 doi:10.1088/1757-899X/1005/1/011001

Preface

These proceedings are a collection of the selected articles presented at the 17th International School-Conference for young scientists and specialists "New Materials", meeting "Advanced technology" (NMAT2019), held in Moscow, Russian Federation on 5 – 8 November 2019.

This conference is regularly organized by *National Research Nuclear University MEPhI* (*Moscow Engineering Physics Institute*), as well as *Scientific and Technical Material Scientists Society* (MSS), Russian Federation. The organizers of this school-conference were also *State Corporation "ROSATOM"*, *National Research Nuclear Center "Kurchatov Institute"*, *National University of Science and Technology MISiS*, *Stock Company "A.A. Bochvar High-technology Research Institute of Inorganic Materials"* and *Research Institute of the Scientific and Production Association (NII NPO "LUCH")*.

The School-Conference was sponsored by the Competitiveness Growth Program of MEPhI ("5–100" Russian Academic Excellence Project - http://5top100.ru).

One of the main goals of this School-Conference was to train young scientists in matters of interdisciplinary scientific communications. Meetings were held in the form of review lectures of the leading scientists with free discussion and personal oral and poster presentations of the participants.

The scope of the conference covered the topics related to the nuclear materials science and promising technologies of the development and processing of new structural and functional materials of future generation of nuclear reactors:

(1) Technology of production and processing of materials; (2) 3D-technology; (3) Nanotechnology; (4) Pulse powder technology; (5) Ion-beam and plasma technologies; (6) Methods for characterization of microstructure and properties of materials; (7) Technologies of ultrafast quenching of melts; (8) Technologies for brazing and welding of reactor components.

The conference was attended by 178 participants from 10 countries (Russia, USA, Belarus, Kazakhstan, Uzbekistan, Azerbaijan, Slovakia, Latvia, Vietnam, Zambia). During the four days of the conference 24 invited lectures, 57 poster presentations were delivered. However, only some papers were qualified for being published in the Open Access IOP Conference Series: Materials Science and Engineering.

Based on the peer-review of reports presented at the school-conference 30 selected papers have been included in this volume. These papers present new research in the various fields of Materials Science and Technology predominantly for Nuclear Engineering. We sincerely hope that the reports presented in this volume will contribute to the advancement of knowledge not only in these fields, but also others.

The organizing committee of the conference thanks all the participants for their fruitful work and personal contribution to the development of these conference proceedings.

Organizing Committee

IOP Conf. Series: Materials Science and Engineering 1005 (2020) 011002 doi:10.1088/1757-899X/1005/1/011002

Peer review declaration

All papers published in this volume of IOP Conference Series: Materials Science and Engineering have been peer reviewed through processes administered by the Editors. Reviews were conducted by expert referees to the professional and scientific standards expected of a proceedings journal published by IOP Publishing.

- **Type of peer review:** Single-blind
- **Conference submission management system:** Conference website and emails
- **Number of submissions received:** 36
- **Number of submissions sent for review:** 36
- **Number of submissions accepted:** 30
- **Acceptance Rate (Number of Submissions Accepted / Number of Submissions Received X 100):** 83.1 %
- **Average number of reviews per paper:** 2
- **Total number of reviewers involved:** 5
- **Any additional info on review process:** MEEDPA 17 (NMET 2019) submissions review process was carried out according to the usual standards of single-blind peer reviewing and in full compliance with the ethical policy of the IOP Publishing.
- **Contact person for queries (please include: name, affiliation, institutional email address):**
Anton Smirnov
National Research Nuclear University MEPhI (Moscow Engineering Physics Institute), 31 Kashirskoe highway, Moscow, Russia
ADSmirnov@mephi.ru

IOP Conf. Series: Materials Science and Engineering 1005 (2020) 012001 doi:10.1088/1757-899X/1005/1/012001

Fabrication processes and testing of the moderator for irradiating devices of BN reactors. Testing of the moderator

K K Polunin and A A Mokrushin

LUCH FSUE, Zheleznodorozhnaya St, 24, Podolsk, Moscow Region, 142103, Russia

E-mail: poluninkk@sialuch.ru

Abstract. The work presents the results of studies of zirconium hydride used as the moderator for irradiating devices in fast reactors. Zirconium hydride fabrication process using methods of direct saturation of the source material and powder preparation and compacting has been optimized. Hydrostatic method was used to measure density of the samples: for the samples produced by direct saturation density varies within 5.61 to 5.65 g/cm^3, for compacted zirconium hydride - 5.45 to 5.55 g/cm^3. Tests to evaluate hydrogen release from zirconium hydride have been conducted. It has been shown that dehydration onset temperature is (420 – 450) °C (powder compaction method) and (500 – 550) °C (direct saturation method). It has been noted that in 20 minutes of the sample heating from 500 to 900 °C, the mass fraction of the hydrogen released is 0.6 %. The results acquired confirm the data on diminished thermal stability of zirconium hydride produced by the powder method in gaseous media.

1. Introduction

Metal hydrides found a wide application in a number of industries as hydrogen sources, reducing agents for production of ultra-pure metals, catalysts, etc. The interest in hydrides in nuclear industry stems from the opportunity to use them as the materials for moderators, reflectors and protective shielding, especially for high-temperature propulsion nuclear reactors [1].

Use of a hydride-based moderator makes possible a significant reduction of the critical mass of the fissile material thereby reducing the dimensions and weight of the reactor itself. To this end, developers of the nuclear power units (NPU) pay close attention to hydrides as prospective moderator materials.

Of all hydrides, zirconium hydride of $ZrH_{1.85}$ is the one most widely used in Russia. Since 1980s it has been employed as “trap-rods” of BOR-60 reactor, with control and protection system rods of similar design being successfully used in BN-350 and BN-600 reactors. Special irradiating devices containing zirconium hydride used for production of gamma-emitters based on ^{60}Co had been developed and are employed in the lateral blanket of the BN-600 reactor [2]. Moderator blocks made of zirconium hydride were used in TOPAZ-1 nuclear power units as components of space vehicles [3]. The HTRE-3 nuclear reactor had been developed in the US to serve as an aircraft propulsion unit [4]. Zirconium hydride was used as the moderator with no hydrogen losses during operation for 100 hours at a temperature of 700 °C. These and other cases confirm the high efficiency and operational capabilities of the products made of zirconium hydride under conditions typical for nuclear reactor cores of various types, especially those operating on fast neutrons.

IOP Conf. Series: Materials Science and Engineering 1005 (2020) 012001 doi:10.1088/1757-899X/1005/1/012001

Zirconium hydride workpieces are usually made by through (direct) saturation of metallic workpieces or by formation of briquettes made of metal hydride powders with their subsequent compacting.

Through saturation or through hydrogenation means diffusive saturation of the metal workpiece to obtain certain hydrogen content in its bulk. An important advantage of the method of small workpieces hydrogenation is the capability to produce the pore-free material and relatively simple production chart. The drawbacks of this method are the considerable time consumption of the workpiece saturation with hydrogen, with processing of a large workpiece requiring up to several weeks and the necessity to employ autoclave equipment in the most cases.

Employment of the method of products formation from metal hydride powders ensures a high production output of the source hydride powders and provides the capability to produce composite materials. The main drawback of the material produced from powders is imperfection of intergranular boundaries resulting in poor resistance to corrosion in gaseous media.

2. Experimental

Production of compact zirconium hydride workpieces was carried out using the “UPS-1” direct synthesis apparatus specially designed for LUCH FSUE [5]. In general, the process of zirconium hydrogenation involves metered supply of hydrogen at a temperature of (600 – 800) °C depending on the required extent of saturation of the material. As the pressure in the ampoule rose to the atmospheric the samples were cooled down to the room temperature in the furnace.

It is worth noting that as the zirconium workpieces get saturated with hydrogen their volume increases noticeably (density of zirconium is 6.5 g/cm^3, density of zirconium hydride in ε-phase is 5.6 g/cm^3). To reduce the scope of the laborious machining of hydride workpieces their dimensions are to be chosen with minimal machining tolerances.

Zirconium hydride workpieces produced by direct saturation were used as the feedstock when the powder compacting method was employed. The workpieces were ground down to a particle size of less than 400 μm using an impact-jet mill. The particle sizes of the powders were monitored using an Analysette 22 laser analyzer of particle size. Subsequently, the method of powder workpieces deformation in steel casings was employed to produce compact products.

Hydrogen content in the samples was measured by high-temperature vacuum extraction using a LECO RH-402 analyzer. The principle of operation of the analyzer is based on high-temperature hydrogen extraction in inert gas flow in a reducing or neutral medium in an induction furnace with temperature programming up to 2800 °C. The extracted hydrogen is then removed by the carrier gas (nitrogen) through the purging system for removal of impurities and transported into a high-sensitivity heat conductivity sensor.

The zirconium hydride samples were weighed using high-precision electronic scales made by VIBRA. Experimental density was measured by hydrostatic method in automatic measurement mode. Besides experimental density, the density of the ideal pore-free material (“theoretical density”) was measured, which was calculated according to reference data and X-ray phase analysis results using the following equation:

$$\rho_T = \frac{A \times z}{N_A} \times \frac{1}{V}, \qquad (1)$$

where A is the atomic mass of all the elements in the compound, a.m.u.; z is the number of formula units per unit cell; N_A is the Avogadro constant, mol^{-1}; V is the unit cell volume, Å^3.

Zirconium hydride samples were studied using a D8 ADVANCE X-ray diffractometer. CuK$_\alpha$ X-rays were used. The smoothest surfaces with suitably large areas were chosen to serve as reflective faces.

Quantitative assessment of porosity of the material was performed using the Vestra Imaging System code for image analysis, using automatic functions of calculation of the volume fraction of the area of the objects (per area unit, %) and report preparation.

Studies of hydrogen release from zirconium hydride were conducted using a Netzsch DSC 404F1 calorimeter and a Netzsch QMS 403 mass-spectrometer. The testing procedure was as follows: the zirconium hydride sample was placed into a ceramic crucible made of Al_2O_3, placed into the high-temperature furnace of the calorimeter. Then the system was flushed with inert gas to remove impurities adsorbed at the surface of the chamber. Then the sample was heated according to the test procedure (heating rate 10 °C/min, maximum testing temperature 1000 °C, isothermal exposure 40 min., cooling simultaneously with the furnace).

Thermogravimetric analysis (TGA) has been performed to make qualitative assessment of hydrogen release from the samples. The measurements were carried out using a Netzsch STA 409 CD synchronized thermoanalyzer with a QMS Aeolos quadruple mass-spectrometer in a high-temperature furnace with a graphite heating element and a protective ceramic tube. The sample was placed into a ceramic crucible made of beryllium oxide and mounted on a TG sample holder on two separating discs made of Al_2O_3. Heating was conducted according to the following scenario:

1) Preheating up to 50 °C at a rate of 5 °C/min, isothermal exposure 5 min;
2) Heating up to 1200 °C at a rate of 20 °C/min;
3) Cooling down to 220 °C at a rate of 20 °C/min.

3. Results and discussion

In accordance with the process operations described above, zirconium hydride samples were fabricated using the direct saturation method and powder production and compacting process. The X-ray diffraction patterns were indicated as belonging to body-centered tetragonal structure (I4/mmm spatial group), corresponding to zirconium hydride ε-phase (the closest to ZrH_2).

Structure of the samples in the cross-section is shown on figure 1. For niobium-doped zirconium hydride, banded substructure of hydride ε-phase is typical. Banding is not very distinctive. The boundaries of the original metallic alloys are clearly visible, because the impurities are precipitated at the grain boundaries. Microstructure of the samples produced by compaction of powders is marked by presence of "small" (up to 30 μm) pores and the extent of porosity varies between 1 and 1.5 %.

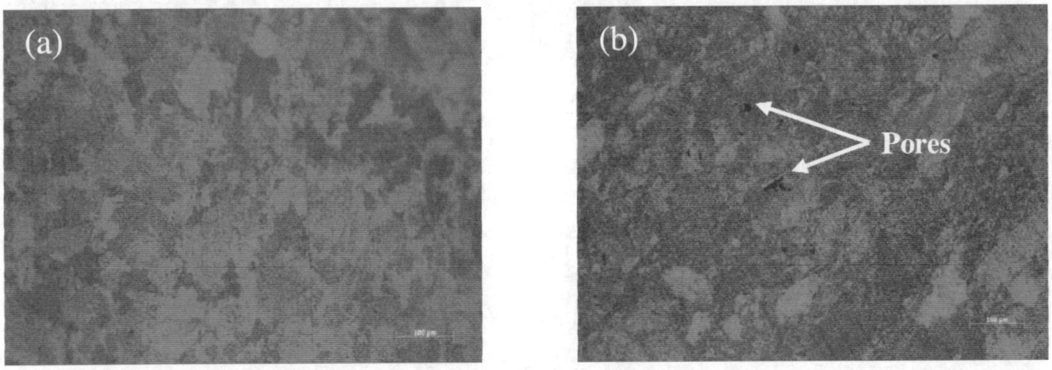

Figure 1. Structure of samples of zirconium hydride produced by direct saturation of material (a) and by powder compaction (b).

According to the results of hydrostatic weighing, the density of the samples produced by direct saturation is (5.61 – 5.65) g/cm^3, and that of compacted zirconium hydride samples is (5.45 – 5.55) g/cm^3, which makes up (98 – 99) % of the theoretical density.

Studies of hydrogen release from zirconium hydride samples fabricated using direct saturation method were conducted in accordance with the procedure described in Section 2. Maximum test temperature was 700 °C, exposure – 40 minutes. This experiment made possible to define the hydration onset temperature which is equal to (500 – 550) °C.

IOP Conf. Series: Materials Science and Engineering 1005 (2020) 012001 doi:10.1088/1757-899X/1005/1/012001

The same tests were conducted for the samples of zirconium hydride produced using the powder compaction process. Here, the maximum test temperature was reduced to 550 °C. Hydrogen release onset temperature was found to be (420 – 450) °C.

To make a qualitative assessment of hydrogen release from zirconium hydride samples, thermogravimetric analysis (TGA) has been conducted. A compact zirconium hydride sample (C(H) = (2.02 – 2.08) %), produced by powder compaction method, was placed into the center of a crucible and heated up according to the temperature scenario described in Section 2 to the maximum temperature equal to 1200 °C. Onset of the mass loss was observed at a temperature of 495 °C (figure 2). Four stages of the mass loss are clearly discernible on the TG-diagram; on the MS-diagram the two first peaks converged and are hard to distinguish. However, it can be said with surety that the mass loss is caused by hydrogen release.

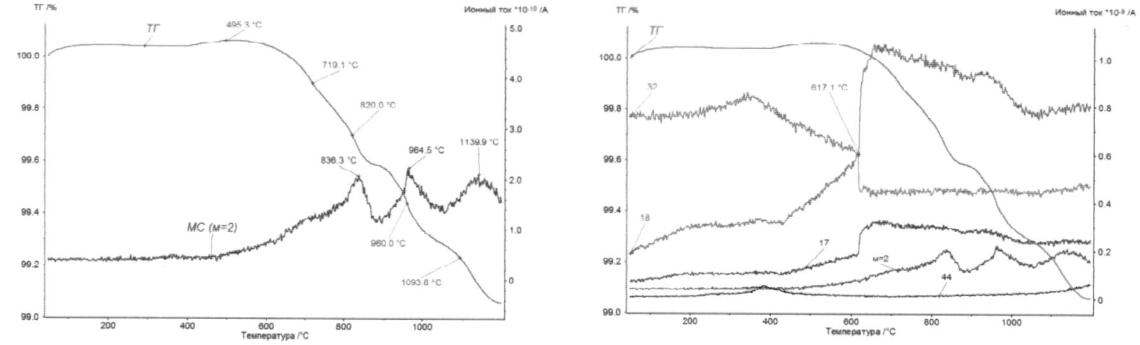

Figure 2. Thermogravimetric (TG) and Mass-spectrometric (MS) diagrams of zirconium hydride decomposition (powder method).

Presence of several peaks on the mass-spectrometric diagram of hydrogen can be due to:

- Occurrence of hydrogen release from the samples in stages when at a given temperature the minimum allowable residual hydrogen concentration in the sample is gained and further hydrogen release may occur only with further elevation of temperature;
- Change of diffusion rate and velocity of hydrogen atom migration towards the surface of zirconium hydride.

Analysis of figure 2 allows to make a quantitative assessment of hydrogen release from the zirconium hydride sample. Throughout the period of (20 – 25) min during sample heating from 500 to 900 °C the weight fraction of the escaped hydrogen reached ~ 0.6 %. As the temperature reached 1100 °C the weight fraction of the hydrogen released from the sample is equal to 1.0 %.

Having analyzed the mass-spectrometric diagrams of zirconium hydride decomposition by oxygen (32), water (18), hydroxide ion (17) and hydrogen (2) the following conclusions can be made:

- Hydrogen release onset temperature (according to the hydrogen diagram) is (495 – 500) °C (the same conclusion was made from the TG-diagram analysis);
- Deviation of the registered parameters from the background value for water and hydroxide ion (OH) was observed somewhat earlier, at a temperature of (420 – 430) °C.

Thus, the hydrogen release onset temperature for the zirconium hydride sample made by powder compaction has been confirmed to be (420 – 430) °C.

The results acquired confirm the assertion that thermal stability of zirconium hydride produced by the powder method is diminished in gaseous media.

4. Conclusion

Zirconium hydride samples have been fabricated by both direct saturation of the original metal and powder production and compaction. Basing on the materials science studies, the characteristics of the samples have been determined (Table 1).

IOP Conf. Series: Materials Science and Engineering 1005 (2020) 012001 doi:10.1088/1757-899X/1005/1/012001

Table 1. Results of materials science studies of zirconium hydride samples.

	Hydrogen weight fraction (%)	Phase composition	Density (g/cm³)	Porosity (%)
Direct saturation	2.02 ± 0.05	ε-phase ZrH_x	5.61 – 5.65	–
Powder compaction	2.07 ± 0.05	ε-phase ZrH_x	5.45 – 5.55	1.0 – 1.5

It has been demonstrated that the method of thermogravimetric analysis of zirconium hydride experimental samples is applicable for quantitative substantiation of behavior of zirconium hydride moderator at operational temperatures ~ 510 °C, as well as during extraction from the core (900 – 1100) °C. The analysis of thermal tests allowed to establish the temperature-time dependencies of hydrogen release up to 1100 °C. It has been shown that ZrH_2 dehydrogenation temperature is (420 – 450) °C (powder compaction method) and (500 – 550) °C (direct saturation method). It was noted that throughout the period of (20 – 25) min during sample heating from 500 to 900 °C the weight fraction of the escaped hydrogen reached ~ 0.6 %. As the temperature reached 1100 °C the weight fraction of the hydrogen released from the sample is equal to 1.0 %.

References

[1] Zaitsev V A, Zaitsev P A and Kozlovsky I V 2019 Hydrides in Nuclear Energy (Moscow: Technosphera) p. 19

[2] Varivtsev A V et al. 2014 *News of Samara Scientific Center of the Russian Academy of Sciences* **16** 112

[3] Pupko V Ya 1996 *Atomic Energy* **80** 357

[4] Karlan A L and Simon G M 1957 *USAEC Report WADC-TM-57-195*

[5] Tolstoukhov S S and Kozlovsky I V 2016 Interaction between Hydrogen Isotopes and Structural Materials. *IHISM'15 Junior: Proceedings of the Tenth International School of Young Scientists and Specialists named after A.A. Kurdyumov,* ed A A Yukhimchuk (Sarov: FSUE "RFNC – VNIIEF") pp 166–70

IOP Conf. Series: Materials Science and Engineering 1005 (2020) 012002 doi:10.1088/1757-899X/1005/1/012002

Study of high temperature oxidation of zirconium fuel element claddings with chromiferous coatings

K K Polunin, A A Mokrushin, A A Urusov and E N Fedorov

LUCH FSUE, Zheleznodorozhnaya St, 24, Podolsk, Moscow Region, *142103,* Russia

E-mail: poluninkk@sialuch.ru

Abstract. This paper presents the results of the study of resistance to corrosion of zirconium fuel element cladding fragments protected with nichrome and chrome coatings. Experiments were conducted on coatings deposition on fuel element claddings by cathodic arc deposition and electrochemical deposition methods. Argon ion beam pretreatment of the cladding surface is shown to have a beneficial effect on coating adhesion. Studies of resilience of the coatings were conducted in the atmosphere of super-heated steam at 900 °C and with an exposure of up to 10000 seconds. It has been shown that the coatings deposited impede formation of ZrO_2 oxide layer at the outer surface and reduce the depth of oxygen penetration into the metallic sublayer as compared with the samples in their original condition.

1. Introduction

Improvement of operational safety of nuclear power units (NPU) is an indispensable condition of atomic energy development. The problems of assurance of nuclear reactors safety and preset fuel burnup are to a large extent determined by operability of fuel assemblies composed of fuel elements, every one of which consisting of zirconium-based alloy cladding and uranium dioxide fuel pellets. In emergency conditions attended by temperature jump due to exothermic zirconium-steam reaction, a violent steam oxidation of the claddings occurs with hydrogen release producing explosive hydrogen-oxygen mixture. Therefore, the problem of enhancing the stability of the nuclear fuel in emergency conditions, i.e. the problem of the accident-tolerant fuel creation arises.

Currently, intensive research and technological development of various methods to enhance safety of light-water reactor fuel element claddings are carried out, ranging from modification of the existing zirconium claddings to development of new fuel element types. Different methods to protect zirconium from corrosion at elevated temperatures are proposed: formation of heat resistant metallic coatings (Al_3Ti, Cr, FeCrAl, NiCr and others) [1 – 3], deposition of protective ceramic coatings resistant to oxidation (Al_2O_3, Cr_3C_2, Si, SiC, SiO_2, Y_2O_3, ZrC, ZrN and others) [4], and modification of the structure and composition of the near-surface layer by ion implantation doping with various elements [5]. Deposition of coatings on zirconium claddings is believed to be the simplest and most optimum method to enhance safety, having virtually no effect on the enrichment of the fuel or the reactor physics, making it possible to anticipate implementation of the new fuel in a relatively short time.

There is a rather wide variety of methods to deposit protective coatings with every one of them having its particular advantages and drawbacks imposing limitations on its scope of application. This paper proposes two methods of coating deposition at the outer surface of fuel element claddings: electrochemical deposition method (galvanic deposition) and cathodic arc deposition (Arc-PVD). The advantages of galvanic deposition are simplicity of process adjustment regarding the thickness of the metal deposited and absence of heating and therefore structural changes of the cladding material [6]. The key factors in favor of cathodic-arc deposition are the high rate of deposition of the coating and the capability to pre-clean the cladding surface with argon ion beam. The advantage of this pretreatment process is preservation of the elemental composition of the near-surface layers and the capability to achieve low surface roughness [7, 8].

2. Experimental

Fuel element claddings with an outer diameter of 9.5 mm made of E110 electrolytic zirconium-based alloy were chosen as the samples for testing.

Preparatory treatment of the samples included cutting into fragments with a length of up to 100 mm, polishing of the butt-ends, surface degreasing with ethanol and subsequent etching in acidic solution carried out to activate the sublayer in order to enhance adhesiveness of the cladding surface.

One of the factors defining quality of the coating deposited and its adhesion to the surface is ion beam pre-cleaning in the chamber immediately before deposition of the coating. To carry out ion beam pre-cleaning the etched samples were placed on a rotary table in a vacuum chamber; with the rotation frequency being (8 – 10) rpm. After the chamber was evacuated to high vacuum, a negative potential was applied to the sample and the sample underwent cleaning by argon ion bombardment for (5 – 10) minutes.

Coating deposition by cathodic-arc method was carried out with a Bulat-6 machine in argon atmosphere under a pressure of (10^{-2} – 10^{-3}) Pa, period of deposition was 120 minutes. To choose the optimum parameters of deposition, a series of experiments was conducted with variation of the shear voltage in the range of 50 V to 450 V. Adjustment of shear voltage applied to the sample made it possible to vary the sample temperature and achieve a high extent of argon ionization. The sample temperature during deposition was measured with a pyrometer.

Electrochemical deposition of a protective coating on fuel element cladding fragments was carried out by immersing the samples into chromium anhydride solution at various operational mode parameters (duration, current density, concentration and temperature of the electrolyte). The sample butt-ends were preliminarily welded with zirconium plugs to prevent ingress of the electrolyte into the interior of the cladding.

To enhance adhesiveness of the coating, diffusive annealing in a vacuum electric resistive furnace was carried out at a temperature of 700 °C for 4 hours. Heating rate of the furnace was ~ 20 °C/min, cooling - together with the furnace itself.

High-temperature corrosion testing of 10 mm long samples was conducted on GAZPAR test bench according to the procedure described in [9]. Testing in superheated steam atmosphere was conducted under stationary conditions in the isothermal zone of the furnace at a temperature of 900 °C with an exposure of up to 10000 seconds. The samples were cooled in a mixed steam-argon atmosphere at a rate of 20 °C/s at most.

Analysis of the microstructure, assessment of the deposited coating thickness and analysis of the elements distribution along the depth of the samples were conducted by electron microprobe analysis using a TESCAN VEGA3 scanning electron microscope.

3. Results and Discussion

Figure 1 presents appearances of the coatings produced at different modes of deposition (100 V, 150 V, 300 V). The appearance of the cladding at 100 V shear voltage didn't change (metallic luster remained); at 300 V shear voltage the cladding surface is oxidated due to its overheating during deposition of the coating (sample temperature reached 650 °C in the course of deposition). The optimal coating was produced at 150 V shear voltage - the cladding became dull, with no trace of spallation and discontinuity, the thickness of the coating made 3 μm (Figure 2a). To make the coating thicker the duration of the process was increased from 120 to 240 minutes making it possible to produce a coating approximately 9 μm thick.

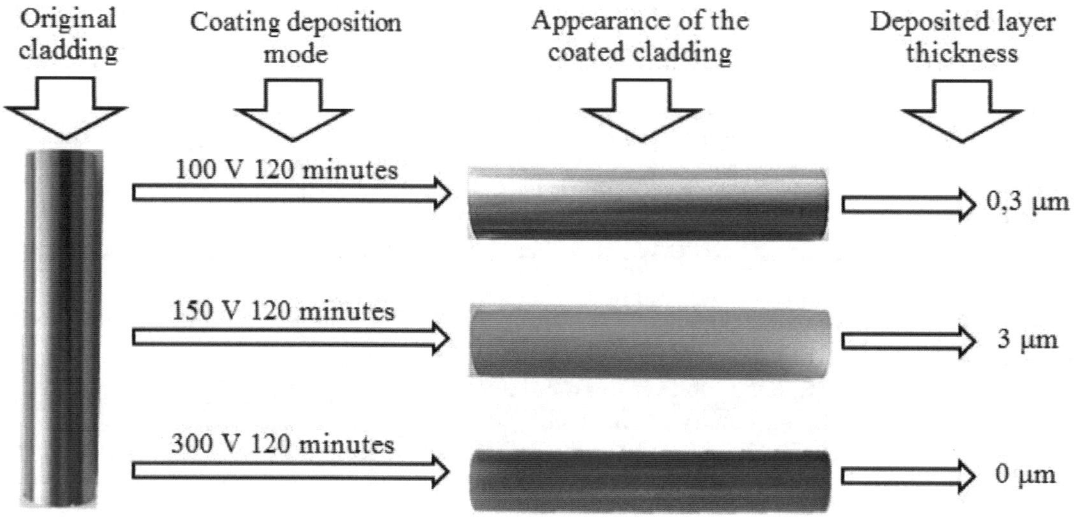

Figure 1. Appearance of the fuel element claddings after deposition of coatings by Arc-PVD method.

Diffusive annealing didn't result in change of appearance of the sample. Electron microscopy of the cladding has shown that after annealing the coating retained its original thickness with no pores, cracks and spallation detected. By the results of the electron microprobe analysis (Figure 2b) the following conclusions have been made: the annealing resulted in active mutual diffusion of elements of the cladding (Zr) and the coating (NiCr), with zirconium having diffused throughout the entire thickness of the coating, whereas in contrast diffusion of nickel into the cladding is insignificant and doesn't exceed (3 – 4) μm; also massive redistribution of chromium towards the outer surface of the coating was observed.

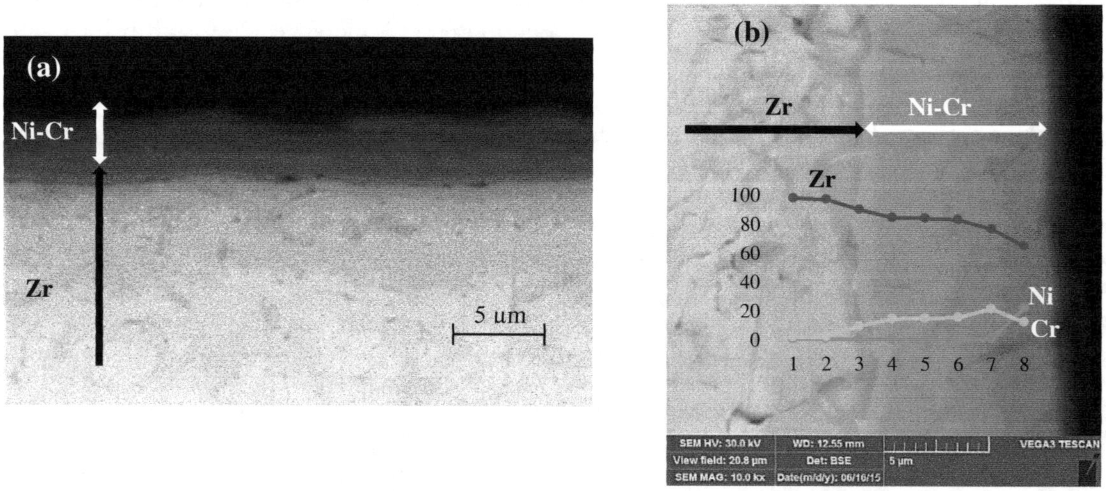

Figure 2. Photographs of the metallographic specimen of the cross section of the cladding.

Since, as a result of the annealing conducted, chromium was redistributed to the outer surface of the cladding and a massive diffusion of zirconium into the coating occurred, a conclusion can be made that the protective part of the coating has significantly thinned, because zirconium diffused into the coating is going to facilitate absorption of oxygen. In this connection a task was set to further increase thickness of the protective coating.

To that end, an additional layer of chromium was deposited on the sample by galvanic method. Rejection of Arc-PVD method is connected with the limitations thereof related to the thickness of the coating deposited and formation of droplet phase in the process of deposition [9]. The coating produced by galvanic method is dull, with no spallation and peeling detected (Figure 3a). The thickness of the deposited layer is approximately 5 μm (Figure 3b).

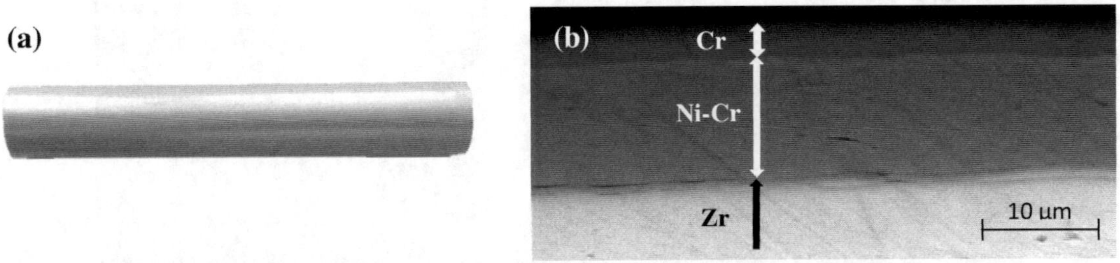

Figure 3. Appearance (a) and cross-section (b) of the fuel element cladding fragment after deposition of the chromium coating layer.

Figure 4 presents the results of high-temperature corrosion testing of the samples at GAZPAR test bench. The testing was carried out on cladding fragments after deposition of NiCr protective coating and subsequent annealing (Figure 4a) as well as on samples with NiCr-Cr coatings (Figure 4b).

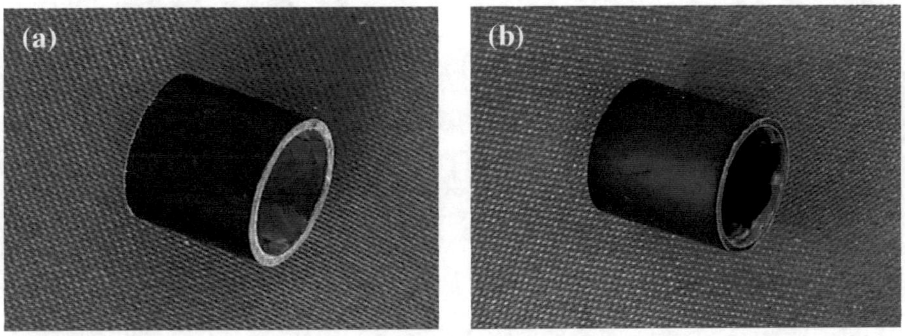

Figure 4. Appearance of fuel element cladding fragments after corrosive testing.

The outer surface of the samples darkened; no cracking, spallation or peeling were found. The inner surface shows traces of spallation of multilayer zirconium oxide (ZrO_2). In general, the appearance of the samples shows no differences.

According to electronic microscopy studies, no formation of zirconium oxide at the outer surface of the sample is observed (Figure 5a, b). The coating retained its thickness, with no cracking and spallation detected. By the results of the electron microprobe analysis (Figure 5c, d) it can be said that a dense layer of Cr_2O_3 oxide has formed on the outer surface of the cladding, preventing penetration of oxygen deep into the coating. Presence of oxygen in the cladding is related with its surface adsorption at the sample surface in the process of sample preparation, therefore, only qualitative assessment of oxygen content in the sample can be made. Presence of internal cracks in the coating is apparently related to formation of brittle (Zr-Ni-Cr) intermetallide.

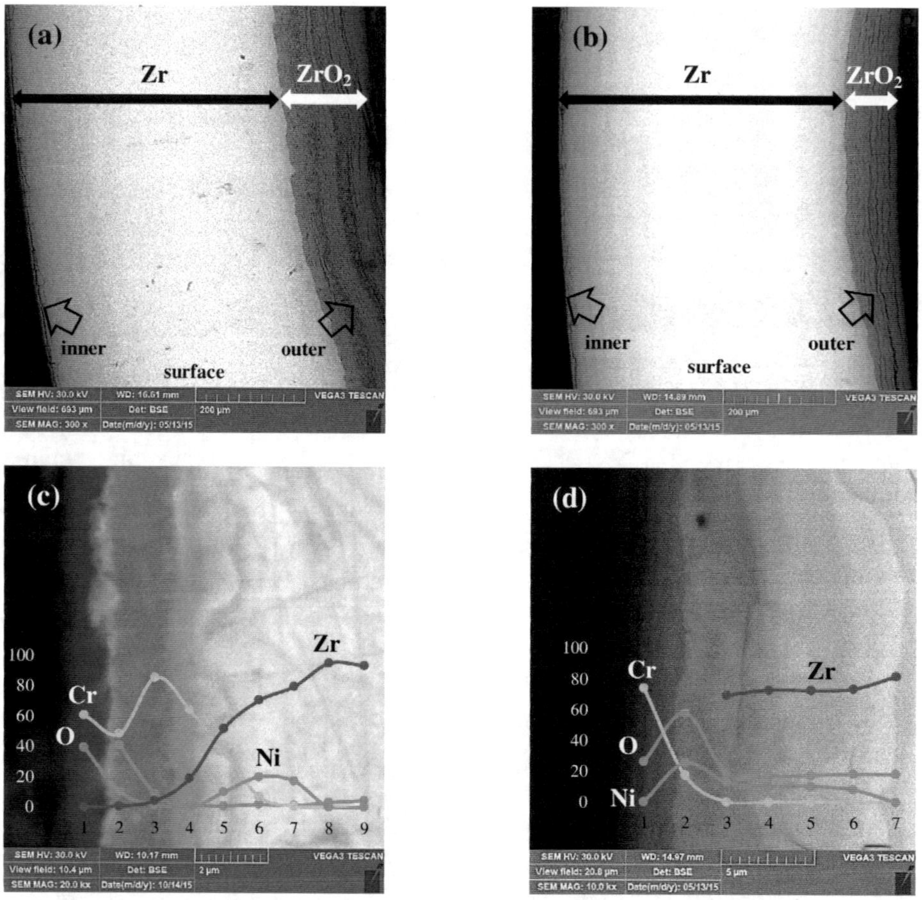

Figure 5. Photographs of the metallographic specimen of the cross section of fuel element cladding fragments after corrosion testing.

4. Conclusion

1. Chromiferous protective coatings have been formed on zirconium cladding. The coatings assure efficient protection of the fuel element claddings (E110 alloy) from corrosive oxidation in superheated steam at T = 900 °C for 10000 seconds.
2. Samples with bi-layer coating (NiCr-Cr) have been made. In the long run, this coating is bound to assure a high degree of protection of fuel elements in emergency conditions with the coolant temperatures as high as 1500 °C.
3. It has been found that diffusive annealing at a temperature of 700 °C for 4 hours results in chromium redistribution towards the outer surface of the cladding, significant diffusion of zirconium along the entire depth of the coating is observed, whereas, on the contrast, nickel diffusion into the cladding is insignificant and makes (3 – 4) μm at most.

References

[1] Kim H-G, Kim I-H, Jung Y-I, Park D-J, Park J-Y and Koo Y-H 2015 *Journal of Nuclear Materials* **465** 531–9
[2] Petelguzov I A 2012 *Problems of Atomic Science and Technology* 2 **78** 114–9
[3] Polunin K K et al 2019 *J. Phys.: Conf. Ser.* **1370** 012050
[4] Ivanova S V, Glagovskiy E M, Khazov I A, Orlov V K, Shlepov I A, Nikitin K N, Dubrovskiy Yu V and Denisov E A 2009 *Physics and Chemistry of Materials Processing* **3** 5–17

[5] Kalin B A, Volkov N V, Valikov R A, Yashin A S and Ignatiyev D N 2017 *IOP Conf. Series: Journal of Physics: Conf. Series* **830** 012085
[6] Andreev A A, Sablev L P and Grigor'ev S N *Vacuum-Arc Devices and Coatings 2010* (Kharkiv: KIPT) p 318
[7] Kalin B A, Volkov N V, Valikov R A and Yashin A S 2017 *Inorganic Materials: Applied Research* **8** 364–9
[8] Yashin A S, Safonov D A, Kalin B A, Fedotov P V and Volkov N V 2017 *IOP Conf. Series: Journal of Physics: Conf. Series* **857** 012057
[9] Polunin K K, Urusov A A, Mokrushin A A, Soldatkin D M, Kuzma-Kichta Yu A, Kiselev D S, Bespechalov B N and Bazyuk S S 2018 *New in Russian Electrical Power Engineering* **9** pp 62–71

Special aspects of uranium oxide non-stoichiometry measurement by thermogravimetric analysis

A Tenishev, R Kuzmin, M Matvenov, V Mikhalchik and D Shornikov

National Research Nuclear University MEPhI (Moscow Engineering Physics Institute) 31 Kashirskoe shosse, 115409, Moscow, Russia

E-mail: matvenov@mail.ru

Abstract. Thermogravimetric analyses for determining uranium dioxide non-stoichiometry have been reviewed. Parameters of oxidation in air at various temperatures have been examined. The key parameters affecting accuracy of the obtained results have been evaluated. Oxidation parameters have been proposed, both for pure uranium dioxide as well as for uranium dioxide with added gadolinium oxide, which makes it possible to determine oxygen to metal (O/M) ratio with at least ±0.002 accuracy in weighed sample portions of approximately 3 g.

1. Introduction
Uranium dioxide is commonly used in modern nuclear reactors as nuclear fuel. Uranium-oxygen compound is one of the most complex systems. It comprises at least ten oxide phases, many of which have several polymorphic forms. For this reason it is important to study the phases occurring in UO_2 oxidation process as well as the effect of external conditions on their formation. This task becomes particularly difficult in case of fuel pellets with alloying additives, for example, burnable neutron absorbers. They are added to inhibit excess reactivity in nuclear reactors at the early fuel operation stage.

2. Materials and Experimental Methods
Oxygen to metal (O/M) ratio is an important parameter in the nuclear oxide fuel technology. This ratio has a direct bearing on chemical activity of fuel in relation to the cladding material as well as to the fission products resulting from nuclear fuel burnup. O/M ratio also has an effect on changes in physical properties, such as thermal conductivity, ductility, melting point, etc. Therefore O/M ratio determination in nuclear fuel is important for development and operation of modern nuclear power reactors.

Thermogravimetric analyses (TGA) are the most universal and accurate as they allow measurements to be performed in a wide non-stoichiometry range. Direct calculation and, consequently, no need of calibration are their undisputable advantages. Proper application of these analyses makes it possible to use small weighed sample portions in any condition – it may be an intact nuclear oxide fuel pellet or a sampled powder material.

Temperature dependence of uranium dioxide oxidation is characterised by a complex behaviour, because intermediate isothermal dwells and final temperature affect both the rate of oxidation as well as the length of incubation period. Dependence of oxidation kinetics on the rate of sample heat-up is low, however, as the heating rate increases incubation period is reported to shorten. The rate of oxidation is highly dependent on sample density – dense samples are more resistant to oxidation, because they have a smaller open porosity and a poorer access of oxygen. Specific powder surface area and grain size also have a strong effect on oxidation of nuclear oxide fuel. The larger the area, the smaller is the grain size, and the faster the process progresses. There is no clearly identified regularity of oxygen partial pressure's effect on oxidation kinetics. In general, gadolinium and soluble additives stabilise fluorite phase which complicates oxidation process. This necessitates selection of optimal temperature parameters in order to study sample composition by TGA.

STA 409 CD (Netzsch, Germany) analyser equipped with Zirconia-M (OOO Issledovatelskiye Tekhnologii, Yekaterinburg) system for monitoring oxygen partial pressure in the furnace atmosphere was used for thermogravimetric studies of nuclear oxide fuel samples. The equipment with solid electrolyte galvanic cell was used to perform an independent evaluation of oxygen potentials as well as non-stoichiometry in the obtained samples. Uranium dioxide sample masses were determined by CPA225D (Sartorius, Germany) scales. JSM – 6610LV (JEOL, Japan) scanning electron microscope with wavelength range and energy-dispersive spectrometers as well as D8 Discover (Bruker, Germany) high-resolution diffractometer were used to analyse sample structures and phase compositions.

Uranium dioxide samples were manufactured using UO_2 powder fabricated by ADU process with the addition of 0.018 at.% aluminium in the form of $Al(OH)_3$.SEM image structure before sintering present in figure 1.

Figure 1. SEM image of the structure of ceramic UO2 with 8% Gd_2O_3 before sintering.

Compacts were manufactured by the process of pressing compacts with the use of aqueous solution of polyvinyl alcohol as binder. Compacts were sintered in static Ar-8%H_2 atmosphere at temperature of 1350 °C for 8 hours. Final density of samples amounted to approximately 90 % of theoretical density (TD). X-ray phase analysis demonstrated the presence of only one phase on UO_2 basis. UO_2 lattice parameter was 547.11±0.04, which corresponds to stoichiometric composition. In addition, oxygen potential (at T = 1000 °C, which amounted to -447±3 kJ/mol) and O/M ratio = 2.0010±0.0005 were determined by means of the equipment with solid electrolyte galvanic cell.

It is common knowledge that UO_2 is oxidised to U_3O_8 in two stages, through forming of U_4O_9 or U_3O_7 oxides, depending on heat treatment parameters [1]. That is why first of all the measurements were performed by differential scanning calorimetry (DSC) combined with thermogravimetry (TG). Heat-up was carried out at the rate of 10 °C/min up to the temperature of 800 °C in the desiccated air with the flow rate of 50 ml/min. Two peaks were identified during oxidation, and the oxidation process terminated at 450 °C. Regarding the peaks, temperature values representing the beginning and the end of reaction were determined: they totalled at 248 and 290 °C for the first peak, and 379 and 394 °C – for the second.

Subsequently, samples were subjected to oxidation in a ceramic crucible made of BeO in the desiccated air with the flow rate of 50 ml/min. For measurement purposes pellets were fragmented to pieces of 1-2 mm, which ensures a larger extended surface contacting with the furnace atmosphere. Samples were heated at the rate of 10 °C/min. The temperature program included two isothermal dwells – at 1200 °C and 1000 °C. Sample mass equalled approximately 3 g. It has been identified that any changes in masses cease at 700 °C. The curve levels off before 900 °C, and with further heat-up to 1200 °C loss in mass is demonstrated. Uranium-oxygen phase diagram [2] implies that at temperatures exceeding 800 °C U_3O_8 already starts to noticeably deviate from its stoichiometric composition. Analysis of oxidised product diffraction patterns showed the presence of α-U_3O_8 phase and traces of β-U_3O_8. Consequently, loss in mass at high temperatures is related to changes in U_3O_{8-x} equilibrium composition.

Therefore, oxidation conditions involving heat-up to 800 °C at the rate of 10 °C/min in the desiccated air with the flow rate of 50 ml/min were selected to determine O/M ratio. Initial O/M ratio in a sample based on oxidation pickup through to stoichiometric U_3O_8 was calculated as per ASTM C1453 [3].

The average O/M value for five measurements amounted to 2.0017±0.0003, which corresponds to the values obtained from X-ray analysis data and by means of the equipment with solid electrolyte galvanic cell. Subsequently, based on the obtained X-ray as well as TGA data, the temperature program for oxidation was modified by adding an intermediate isothermal dwell at 600 °C.

For the purposes of checking the proposed procedure the samples with over-stoichiometric composition were obtained both by specifying oxygen potential of furnace atmosphere as well as by controlled oxidation in air. O/M ratios for these samples, determined by means of the solid electrolyte galvanic cell technique stood at 2.010±0.003, 2.020±0.003, and 2.050±0.005. The following O/M ratios were obtained by TGA: 2.011±0.001, 2.019±0.001, and 2.054±0.001. Thus, a good agreement of results obtained by different methods was achieved, which indicates a sufficient efficiency of the procedure for over-stoichiometric samples of undoped uranium dioxide.

The samples with 8 wt.% Gd_2O_3 content were manufactured by the process using dry lubricant (binder) – aluminium stearate (estimated Al quantity amounts to 0.023 at.%). Sintering was performed in a dilatometer with dwell time of about 8 hours at maximum temperature of 1600 °C in Ar-8% H_2 atmosphere. Final density stood at approximately 10.3±0.1 g/cm^3 (96% TD). SEM image structure after sintering is present in figure 2.

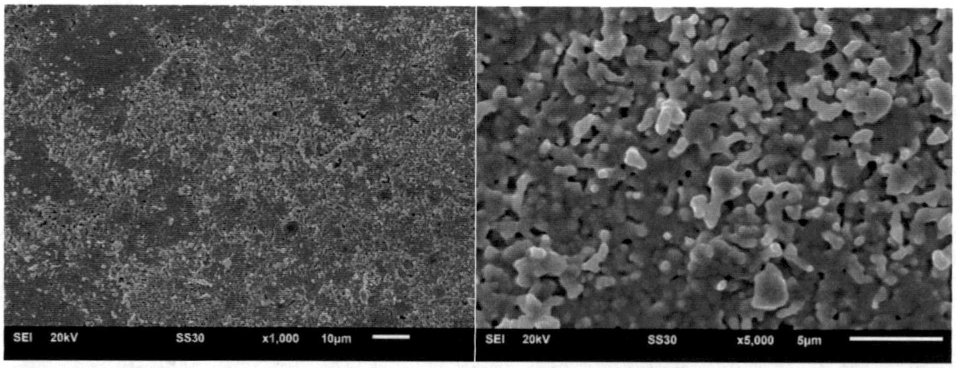

Figure 2. SEM image of the structure of ceramic UO_2 with 8% Gd_2O_3 after sintering.

The conducted X-ray analysis showed that only one $(U_{1-y}Gd_y)O_{2\pm x}$ phase was present, with lattice parameter of 545.97±0.04 pm.

The papers dedicated to studying oxidation of uranium dioxide doped with rare earth metals [4-6] report an increased stability of fluorite-type structure as compared to U_3O_8 (at alloying element concentration of 4 to 10 wt.%). It is shown in papers [7, 8] that when performing oxidation at temperatures exceeding 350 °C in air an unstable $(U,Gd)_3O_8$ phase may be obtained. Within 350-450 °C range samples may demonstrate the presence of U_4O_9, U_3O_8 phases as well as a strongly over-stoichiometric UO_{2+x} with Gd dissolved in them, and at temperature of 450 °C and higher only a solid gadolinium solution is observed in U_3O_8. According to data provided in [7, 9], at temperatures exceeding 950 °C phases get separated into pure U_3O_8 and $(U,Gd)O_X$.

Measurements according to the temperature program involving heat-up to 1200 °C at the rate of 10 °C/min were performed to study oxidation kinetics in detail. Intense increase in mass already ceases at 590 °C, concurrently, the change in the rate of oxidation is observed at around 450 °C. Then the mass continues to slightly grow up to 856 °C, after which it starts to decrease through to 1200 °C. Loss in mass appears to be related to the separation of $(U_{1-y}Gd_y)_3O_8$ solid solution. It is of interest to note that ASTM C968 [10] recommends the maximum oxidation temperature of 900 °C, which is close enough to the temperature corresponding to the start of loss in mass. Therefore the primary task when selecting oxidation conditions for samples with gadolinium is to ensure the final composition of $(U,Gd)_3O_8$, because O/M ratio is calculated as per [10] exactly according to this assumption.

Further experiments conducted in combination with X-ray phase analysis of oxidation products showed that optimal conditions involve heat-up to 475 °C at the rate of 20 °C/min, dwell time of 4 hours, and cooling together with the furnace. After this the only phase present in the final product is $(U,Gd)_3O_8$. Initial O/M ratio in $(U_{1-y}Gd_y)O_{2\pm x}$ samples is substoichiometric and stands at 1.988±0.002, which conforms to the sintering conditions in hydrogen-containing atmospheres. The main contributing components of the error in determining O/M ratio in this case are the selection of the required mass of the weighed sample portion for accurate pickup determination as well as the selection of oxidation conditions that would ensure that $(U_{1-x}Gd_x)_3O_8$ of stoichiometric composition is obtained as final product.

3. Conclusions

Temperature programs have been proposed to perform thermogravimetric measurements of non-stoichiometry in the samples of pure uranium dioxide or uranium dioxide doped with gadolinium. For the pure uranium dioxide the best results was obtained after heat-up to 800 °C at the rate of 10 °C/min in the desiccated air with the flow rate of 50 ml/min with an intermediate isothermal dwell at 600 °C. For $(U,Gd)O_2$ samples optimal conditions is a heat-up to 475 °C at the rate of 20 °C/min, dwell time of 4 hours, and cooling together with the furnace. The obtained parameters have been checked on samples with the known oxygen content measured by other methods. Accuracy and reproducibility of obtained data have been shown as good. The error of the selected O/M ratio measurement conditions does not exceed ±0.002.

Acknowledgements

This work was supported by Competitiveness Growth Program of the Federal Autonomous Educational Institution of Higher Education National Research Nuclear University MEPhI (Moscow Engineering Physics Institute).

References

[1] Valdivieso F, Francon V, Byasson F, Pijolat M, Feugier A and Peres V 2006 Oxidation behavior of unirradiated sintered UO_2 pellets and powder at different oxygen partial pressures, above 350 °C *Journal of Nuclear Materials* **354** 85–93

[2] Caneiro A and Abriata J. 1984 Equilibrium oxygen partial pressure and phase diagram of the uranium-oxygen system in the composition range 2.61 < O/U < 2.67 between 844 and 1371 K *Journal of Nuclear Materials* **126** 255–67

[3] ASTM C1453-00. Standard Test Method for the Determination of Uranium by Ignition and the Oxygen to Uranium (O/U) Atomic Ratio of Nuclear Grade Uranium Dioxide Powders and Pellets

IOP Conf. Series: Materials Science and Engineering 1005 (2020) 012003 doi:10.1088/1757-899X/1005/1/012003

[4] Hanson B 1998 The burnup dependence of light water reactor spent fuel oxidation. PNNL-11929 Report (Pacific Northwest Laboratory, Richland, Washington)

[5] Thomas L, Einziger R and Buchanan H 1993 Effect of fission products on air-oxidation of LWR spent fuel *Journal of Nuclear Materials* **201** 310–9.

[6] Campbell T, Gilbert E, White G, Piepel G and Wrona B 1989 Oxidation behavior of non-irradiated UO_2 *Nuclear Technology* **85** 160–71

[7] McEachern R and Taylor P 1998 A review of the oxidation of uranium dioxide at temperatures below 400°C *Journal of Nuclear Materials Volume* **254** (2–3) 87–121

[8] You Gil-Sung, Kim Keon-Sik , Min Duck-Kee and Ro Seung-Gy 2000 Oxidation kinetic changes of UO_2 by additive addition and irradiation *Journal of Nuclear Materials* **277** (2–3) 325–32

[9] Yang H, Kang K W, Kim K S, Song K W and Kim J H 2001 The high temperature oxidation behavior of 10wt.% doped UO_2 *Journal of Korean Nuclear Society* **33** 307

[10] ASTM Designation C968-99 Standard Test Methods for Analysis of Sintered Gadolinium Oxide-Uranium Dioxide Pellets

IOP Conf. Series: Materials Science and Engineering 1005 (2020) 012004 doi:10.1088/1757-899X/1005/1/012004

Radiation effects in coatings ZrTiN induced by impact of high energy krypton ions

T Adabergenova[1,2], S Kislitsin[2,3], V Uglov[3,4], I Ivanov[2], A Larionov[2] and A Ryskulov[2]

[1] Al-Farabi Kazakh National University, Almaty, Kazakhstan
[2] RSE INP Republic of Kazakhstan, Almaty, Republic of Kazakhstan
[3] NRNU MEPhI, Moscow, Russia
[4] Belarus State University, Minsk, Belarus

E-mail: tamaraalga@mail.ru

Abstract. This paper presents research results of the impact of high-energy krypton ions on structure and mechanical properties of ZrTiN coatings. The ZrTiN coatings were synthesized by the magnetron sputtering method on the substrates from Cr18Ni10Ti steel. Samples irradiated by $^{84}Kr^{14}$ with energy 125 MeV to fluence 10^{16} cm^{-2} at room and 600°C temperatures. It established that irradiation resulted in significant changes in the surface structure, i.e. appearance of the coral-like structure. Formation of coral-like structure leads to degradation of mechanical properties such as hardness and corrosion resistance. Most significant changes surface structure and properties of coating observed after irradiation at room temperature.

1. Introduction

One of the promising directions for the development of radiation-resistant materials for nuclear and thermonuclear installations is the use of multicomponent coatings on traditional materials used with properties unattainable for pure materials or alloys. For applying such coatings, advanced technologies, such as magnetron sputtering, vacuum condensation, etc., are usually used [1]. It can be ceramic coatings, as well as coatings based on transition metal nitrides. In particular, coatings based on transition metal nitrides have demonstrated good properties [2-3]. The series of works [4-6] are devoted to the synthesis of ZrTiN coatings by magnetron and vacuum-arc methods with different contents of zirconium and titanium on different substrates (silicon, steel) and the study of their structure and properties. It was shown that these coatings have a good set of properties — high strength characteristics, corrosion resistance, electrical conductivity, etc. Studies of the radiation resistance of ZrTiN coatings showed high stability of the structure and properties when irradiated with heavy low-energy ions and alpha particles [7–8]. This work is a continuation of studies of the stability of the structure and properties of ZrTiN coatings [7] as applied to irradiation with high-energy heavy ions (Kr).

2. Materials and experimental methods

To study the effect of ion irradiation on the structure and properties of the ZrTiN coatings, were prepared the samples with the ZrTiN coatings on substrates of Cr18Ni10Ti structural steel . Since the coatings with the content of Zr and Ti of 50/50 at.% [7] showed the highest stability of properties under irradiation with low-energy heavy ions, we synthesized the coatings with such ratio of Zr and Ti in the coating.

The procedure of sample preparation can be divided on two stages. The first stage was to prepare the substrate for coating application. The substrates of 20 mm × 20 mm × 1 mm dimensions were cut from a massive plate of stainless steel Cr18Ni10Ti by the method of electro-erosion cutting, followed by the standard preparation procedure for coating application: mechanical grinding with a gradual reduction in the particle size of abrasive paper and polishing on cloth using GOI paste. Before coating, the substrates were cleaned from contamination by ion etching with low-energy argon ions.

At the second stage ZrTiN coatings were synthesized by the method of magnetron sputtering of targets from two magnetrons. The sputtered targets were the discs of Zr (99.96 wt.% purity) and Ti (99.99 wt.% purity). The gas mixture with nitrogen content of ~ 25% was used as a plasma-forming gas. The thickness of the applied coatings was controlled by the coating time. As a result, the samples with the ZrTiN coatings ~600 - 700 nm thick were produced.

Thickness of the coatings was checked by the Rutherford backscattering (RBS) method on protons at the accelerator UKP of the Institute of Nuclear Physics [9]. According to the results of processing the spectra of backscattered protons, it was established that the thickness of the ZrTiN coating is 780 nm, and the elemental composition as follow: Zr is 30 at.%, Ti is 31 at.% and N is 39 at.%.

The studies of the structure and elemental composition of the coatings were performed by X-ray structural analysis (XSA), scanning electron microscopy (SEM), energy dispersive analysis (EDS) and atomic force microscopy (AFM).

According to XSA, the initial, before irradiation, structure of the ZrTiN coating consists of two-phase: main phase is a solid solution of ZrN in TiN. Lattice parameter of the solid solution is 4.377 Å. Also, a phase of pure titanium with the lattice parameter of 4.0357Å is observed in minor amounts in the coatings. The diffraction patterns of the steel substrate also show the reflections.

The surface structure of the ZrTiN coating revealed by SEM and AFM methods is shown in figure 1 a,b. We can see from images on the figure 1 that the surface structure follows the structure of the steel substrate - there are bands characteristic of the surface after grinding before coating application.

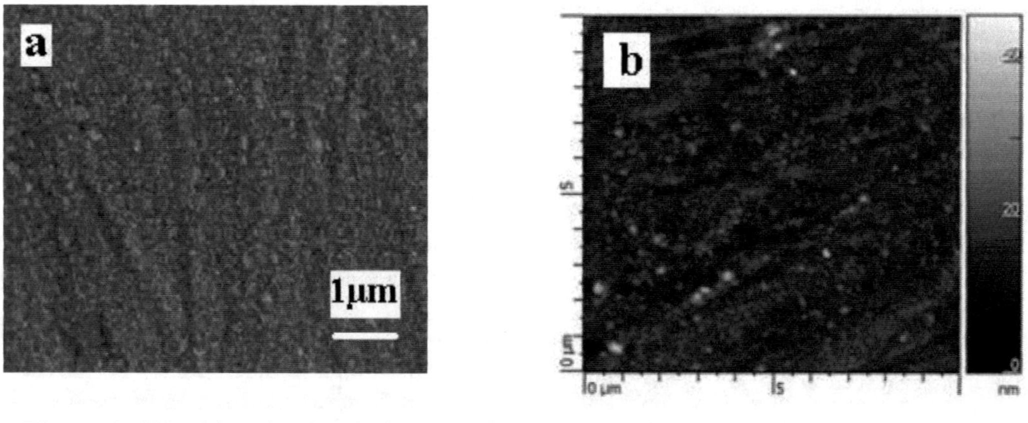

Figure 1. SEM (a) and AFM (b) images of the ZrTiN coating surface before irradiation.

The coating surface has a characteristic relief, which is a finely dispersed globular-shaped particle with an average diameter of ~ 0.4 μm (~ 75% of the surface). The minimum particle diameter is ~ 0.15 μm (~ 20%), the maximum is 0.8 μm (~ 5%).

The results of determining the elemental composition of the ZrTiN coating by EDS method are shown in figure 2.

Figure 2. Distribution of Zr, Ti, and N elements in the coating according to EDS analysis.

The figure 2 shows that zirconium, titanium and nitrogen are evenly distributed over the surface of the coating.

The physical-mechanical properties were determined by hardness (Hn) of the coatings measured on the nanohardness meter "Nanoscan Compact" and the corrosion resistance (CR) in a 3% NaCl solution using the "Gamry Instruments" corrosion cell.

The measured hardness of the ZrTiN coating (on surface) before ion irradiation was Hn = 10.8 GPa. We shall note, that hardness of the ZrTiN coating is more than 2 times higher than hardness of steel (4.4 GPa). The corrosion rate of the ZrTiN coating was - 12.3×10^{-3} mm/year. The results of hardness and corrosion resistance measurements are also shown below in Table 2.

3. Irradiation procedure

The ZrTiN samples were irradiated with high-energy krypton ions on the high-energy channel of the heavy ion accelerator DC-60 at the Institute of Nuclear Physics branch in Nursultan city. Irradiation at room temperature means that was no special heating of specimen during irradiation. Real temperature at the surface of specimen under action of ion beam does not exceed 100°C. The target holder with electric heating was used to irradiate the sample at 600 °C temperature. The parameters of ion beam and conditions of irradiation with high-energy krypton ions are shown in Table 1.

Table 1. Irradiation conditions for the samples with ZrTiN coatings by high-energy krypton ions

	Ion	Energy, MeV	Irradiation temperature	Fluence, cm^{-2}	Beam current, μA	Irradiation area, cm^2
ZrTiN	$^{84}Kr^{14+}$	125	Room	10^{16}	0.35	4
ZrTiN	$^{84}Kr^{14+}$	125	600°C	10^{16}	0.21	3.01

4. Experimental results and discussion

Surface sputtering of the ZrTiN coating under high-energy krypton ions bombardment is minimal. Thickness measurement of the coating, irradiated with 125 MeV krypton ions by the POP method showed that, within the accuracy of the method, the coating thickness did not change for both the sample irradiated at room temperature and the sample irradiated at 600 °C.

The phase structure of the ZrTiN coating did not change both after irradiation at room temperature and at a temperature of 600 °C. On diffractograms observed reflections the phase of solid solution ZrN in TiN, and phase of pure titanium. Shift of X-ray peaks allowed revealing increase in the lattice parameters of both phases. An increase in the lattice parameter is caused by the formation of radiation defects upon irradiation with krypton ions. It should be noted that irradiation at a temperature of 600 °C leds to a larger increase in the lattice parameter of phases both solid solution and pure titanium compare with irradiation at room temperature.

No changes in the ratio of Zr and Ti concentrations in the samples irradiated at room temperature and 600 °C were detected by EDS. Zirconium, titanium and nitrogen are evenly distributed over the surface of the coating as before irradiation, see figure 2.

The surface morphology (SEM and AFM images) of the ZrTiN coating irradiated with 125 MeV Kr ions to the fluence of 10^{16} cm^{-2} at room temperature is shown in figure 3 a, b. From these images can see that the surface structure has changed drastically: on the irradiated surface of the ZrTiN coating is observed formation of the coral-like structure. At high magnifications, it can be seen that the surface is covered with rounded and oblong hillocks. Rounded-shaped hillocks are traces from single incident heavy ions, and oblong is overlapping traces from two or more incident heavy ions. Hillocks appearance is caused by the energy release at braking of Kr ions in the surface layers of the coating. As already mentioned above, the appearance of hillocks does not accompanied in changes of coating elemental composition.

Figure 3c, d shows the structure of the ZrTiN coating surface irradiated with the same energy Kr ions and up to the same fluence, but at the temperature of 600 °C.

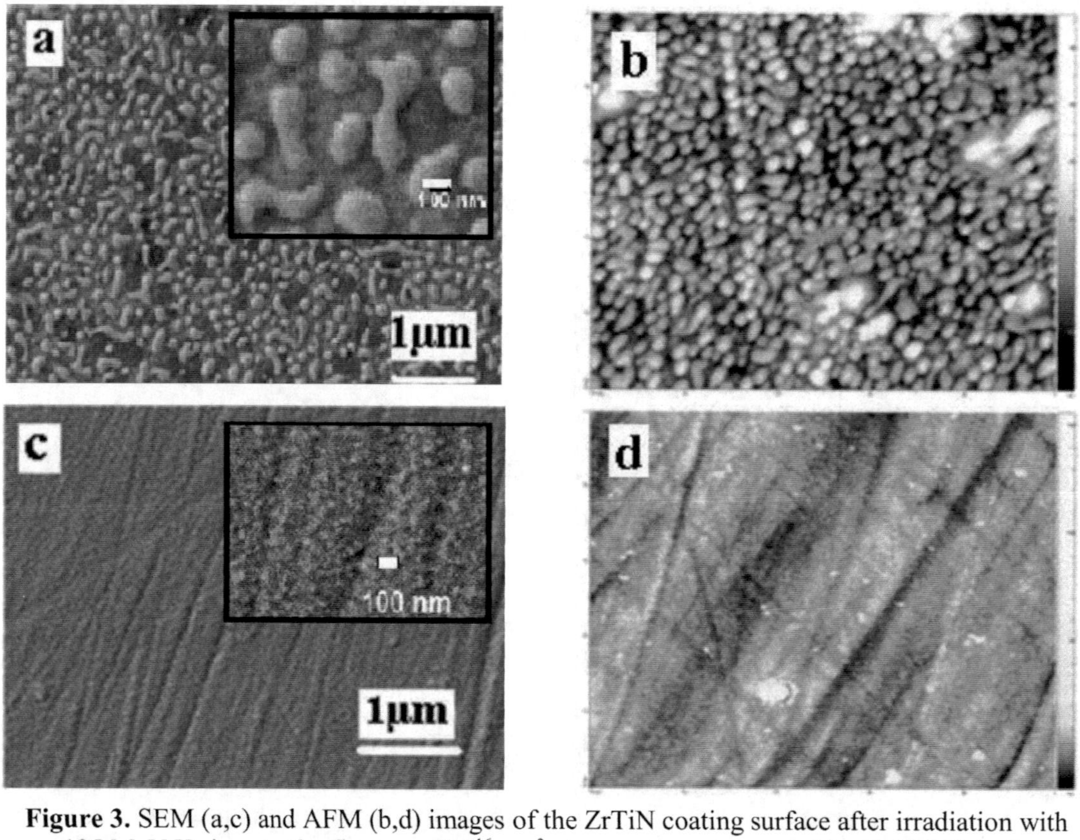

Figure 3. SEM (a,c) and AFM (b,d) images of the ZrTiN coating surface after irradiation with 125 MeV Kr ions to the fluence of 10^{16} cm^{-2} at room temperature (a,b) and 600 C (c,d).

Irradiation at 600 °C to a much lesser extent affects the change in surface structure. Small hillocks ~5-8 nm in size are observed on the irradiated surface, which, apparently, also represents the traces of incident Kr ions. Apparently, due to the relatively high irradiation temperature, the damage region from the incident Kr ion partially anneal, and, as a result, the hillocks on the surface are much smaller in size compared to irradiation at room temperature.

Changes in physical and mechanical properties (hardness and corrosion resistance) correlate with the changes in surface structure. The measurement results are shown in Table 2. It follows from the Table 2 that the decrease in hardness of the surface, irradiated at room temperature is more significant compared with hardness of the surface, irradiated at 600 °C. A similar dependence is observed for corrosion resistance – corrosion resistance degrades to a greater extent after irradiation at room temperature.

Table 2. Hardness and corrosion resistance of the ZrTiN coatings before and after irradiation with high-energy krypton ions.

Non-irradiated		Irradiated with 125 MeV Kr ions			
Hardness GPa	Vcorr, mm/year	Fluence, cm^{-2}	Irradiation temperature	Hardness, GPa	Vcorr, mm /year
10.8	12.3×10^{-3}	10^{16}	room	5.8	75×10^{-3}
			600 °C	9.3	24×10^{-3}

5. Conclusion

The article presents the results of experimental studies of the effect of irradiation with 125 MeV Kr ions on the structure and properties of thin (~ 800 nm) ZrTiN coatings. It is established:

- Irradiation of the ZrTiN coating with high-energy Kr ions at room and 600 °C temperatures does not lead to changes in the structural-phase composition of the coating and redistribution of the coating components;

- Irradiation of the ZrTiN coating with 125 MeV by Kr ions led to significant changes in the surface structure – the appearance of a coral like structure. Rounded and oblong hilloks formed on the surface at the sites of interaction of incident ions with the coating surface. Irradiation at room temperature causes more significant changes in the surface structure compared with irradiation at elevated (600 °C) temperatures.

- Irradiation of the ZrTiN coating with high-energy krypton ions leads to a decrease in the hardness and corrosion resistance of the coating. Irradiation at low temperature leads to greater degradation of physical and mechanical properties.

Acknowledgements
This work was performed in the framework of the program "Scientific and technical support for experimental research at the Kazakhstan materials science tokamak ktm" of the Ministry of Energy of the Republic of Kazakhstan

References

[1] Murty K L and Charit I 2008 *J. Nuclear Materials* **383** 189–195
[2] Ciriello A, Rondinella V V, Staicu D and Somers J *J. Nuclear Materials* 2007 **371** 129–133
[3] Gavarini S, Toulhoat N, Peaucelle C, Martin P, Mende J, Pipon Y and Jaffrezic H 2007 *J. Nuclear Materials* **362** (2–3) 364–373
[4] Uglov V V, Anishchik V M, Zlotski S V, Abadias G and Dub S N 2008 *Surface and Coatings Technology* **202** (11) 2394–98
[5] Uglov V V, Anishchik V M, Zlotski S V and Abadias G 2006 *Surface and Coatings Technology* **200** (22–23) 6389–94
[6] Saladukhin I A, Abadias G, Michel A, Uglov V V, Zlotski S V and Dub S N 2015 *Thin Solid Films* **581** 25–31.
[7] Uglov V V, Rusalski D P, Zlotski S V, Sevriuk A V, Abadias G, Kislitsin S B, Kadyrzhanov K K, Gorlachev I D and Dub S N 2010 *Surface & Coatings Technologies* **204** 2095–98
[8] Van Vuuren A J, Skuratov V A, Uglov V V, Neethling J H and Zlotski S V 2013 *J. Nuclear Materials* **442** 507–511
[9] Arzumanov A A, Borisenko A N, Gorlachev I D, Eliseev A S, Kadyrzhanov K K, Lysukhin S N, Platov A V and Sapozhnikov A 2002 Tandem accelerator UKP-2-1 in nuclear-physical investigations *Preprint N 21* (Almaty: Institute of nuclear physics) p 35
[10] Kislitsin S, Gorlachev I and Uglov V 2015 *Acta Physica Polonica A* **128** 818–823
[11] Kislitsin S B, Uglov V V, Gorlachev I D, Ivanov I A and Larionov A S *Materials of the 11th International Conference* (Minsk: Minsk BSU publishing house) pp 355–357

IOP Conf. Series: Materials Science and Engineering 1005 (2020) 012005 doi:10.1088/1757-899X/1005/1/012005

Ion implantation of helium and hydrogen in boron coatings

A N Karpikov[1], A S Larionov[1], S B Kislitsin[1,2], I I Chernov[2], M S Staltsov[2], A S Dikov[1] and S O AkaeV[1,3]

[1] Institute of Nuclear Physics of the Republic of Kazakhstan, Ibragimov st. 1, 050032 Almaty, The Republic of Kazakhstan
[2] National Research Nuclear University MEPhI (Moscow Engineering Physics Institute), Kashirskoe highway 31, 115409 Moscow, Russia
[3] Satbayev University, Satpaev st. 22a, 050013 Almaty, The Republic of Kazakhstan

E-mail: larionov-inp@yandex.kz, i_chernov@mail.ru

Abstract. The work presents the calculated mean free paths of hydrogen, deuterium and helium ions in boron-carbon and boron-titanium films of various configurations. Has been rated impact of these coatings at various film thicknesses for ion capture in tungsten. The calculations were carried out using the software package SRIM-2012, for each mileage value was calculated by modeling of10,000 cascades.

1. Introduction

One of the critical important problem in thermonuclear research is the protection of plasma-facing components from surface radiation damage. The first wall of the tokamak chambers must withstand loads of very high power (several MW/m^2), high thermal stresses, and should not strongly pollute the plasma with impurities of elements with large atomic numbers [1, 2]. These impurities appear as a result of the interaction of the plasma with the first wall. In addition to radiation resistance, the capture and recirculation of hydrogen isotopes is also taken into account [3]. Similar requirements are also imposed on other reactor components that come into contact with plasma: limiters, plate targets in magnetic divertors, and antennas for auxiliary plasma heating.

The creation of materials that do not pollute the plasma is of fundamental importance to the physics of fusion. More suitable are surface-radiation-resistant materials with a low serial number: Be, Al, Ti, V and etc. as well as Li, B, C are also applies [4]. The compounds of these elements were investigated from the point of view of creating radiation-resistant coatings [5–8].

The harmful effects of heavy types of atomic nuclei appearing in the hydrogen plasma as impurities have already been noted in the early stages of thermonuclear research. The most dangerous phenomenon leading to plasma pollution is the cathodic sputtering of the wall material under the action of bombardment by flows of ions and atoms of deuterium and tritium, alpha particles and neutrons, as well as impurity ions and atoms.

For protection, the practice of boronization of the chamber walls is widely used [9]. The boron – carbon coatings obtained in this method reduce sputtering and affect the capture of hydrogen isotope ions by materials of the first wall and divertor. Nevertheless, in the process of work, they are quickly destroyed and lose their protective properties [10].

In this work it has been suggested that titanium boride-based coatings may have a longer life with protective properties similar to boron-carbon films. The resistance of titanium boride to ion sputtering

IOP Conf. Series: Materials Science and Engineering 1005 (2020) 012005 doi:10.1088/1757-899X/1005/1/012005

was studied by us earlier in [11]. Also, the previously obtained results showed sufficient efficiency of using the DC-magnetron sputtering method of a composite cathode target to obtain coatings with the boron-titanium system [12].

2. Materials and methods

The boronization process is carried out in situ. The procedure for the deposition of boron – carbon films is constantly being improved; the difference in the methods used leads to different B/C ratios. The difference in the composition of coatings depending on the methods which are used in the synthesis plants, it is possible to trace in [13–15]. In this paper, we take the average values of the most common configurations of such films. The parameters of boron – titanium coatings correspond to a promising neutron absorbing titanium boride coating with a high boron content [12]. The elemental composition of the compounds is presented in Table 1. Coating No. 1–3 are B-C compounds with different B/C ratios; coatings No. 4–5 are B-Ti compounds with different B/Ti ratios. Pure tungsten is considered as a substrate as the main candidate of plasma-facing materials for structural elements.

Table 1. The elemental composition of the studied coatings

No.	B	C	H	O	Ti
1	41.6	8.3	50		–
2	60.48	27.11		11	–
3	86	6	8		–
4	52	12.5	–	26.8	8.7
5	76	4	–	4	14

Simulation of ion irradiation processes were carried out using the software package SRIM-2012, for each mileage value is calculated by modeling 10,000 cascades. The mean free paths of H^+, D^+, and He ions are calculated. The ion energy values characteristic of the fusion reaction were used. The energy of the isotopes of hydrogen is 14.8 MeV, helium is 3.25 MeV. The results obtained made it possible to evaluate the effect of coating parameters on capture of H^+, D^+, and He^+ ions in the tungsten.

3. Results and discussion

Coatings with high boron content are sufficiently transparent for the flow of high-energy ions of hydrogen and helium isotopes. This property helps to avoid intensive spraying of the coating. The performed model calculations confirmed this fact (Table 2).

Table 2. The sputtering coefficient of tungsten and coatings 1–5, at./ion

	W	1	2	3	4	5
H^+	0	0	0	0	0	0
D^+	0	0	0	0	0	0
He^+	8×10^{-4}	10^{-4}	2×10^{-4}	4×10^{-3}	9×10^{-4}	4×10^{-4}

But at the same time, the passage of ions in the material leads to the appearance of cascades of collisions, leading to the redistribution of elements. In Figure 1a shows the distribution of boron-containing defects in coating 1 (thickness is 200 nm) and defects containing tungsten in the substrate. Defects in the coating formed at a depth of > 150 μm. The highest concentration of these defects is concentrated near the boundary with the substrate. A small amount of boron transferred in the substrate. In tungsten, a larger number of defects appear along the entire path of the passage of H^+ ions. The main part of the displaced W atoms is concentrated at a depth of ~ 550 μm. The distribution of H^+ ions is similar to the distribution of W atoms (Figure 1b).

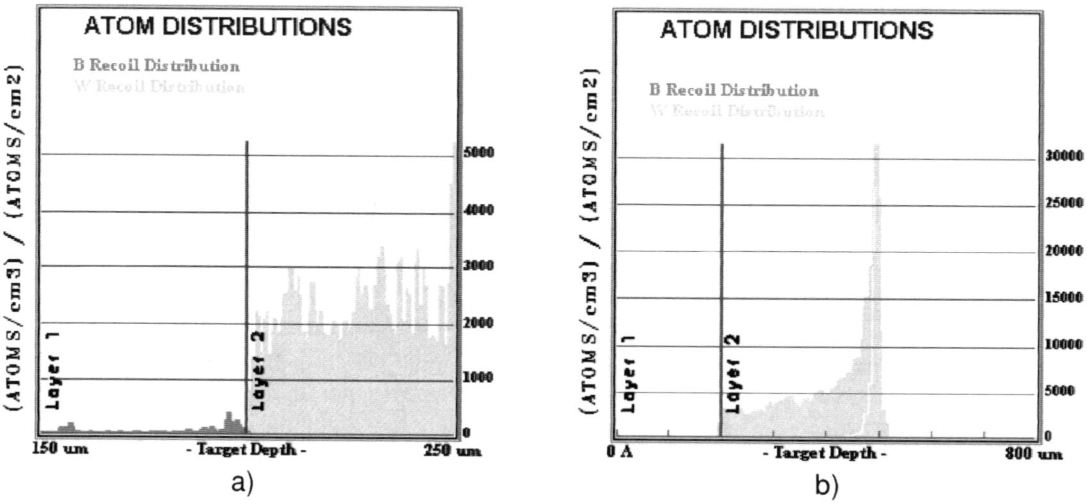

Figure 1. The simulation results of irradiation by H^+ ions of coating 1 on tungsten: (a) redistribution of B and W; (b) distribution of H^+.

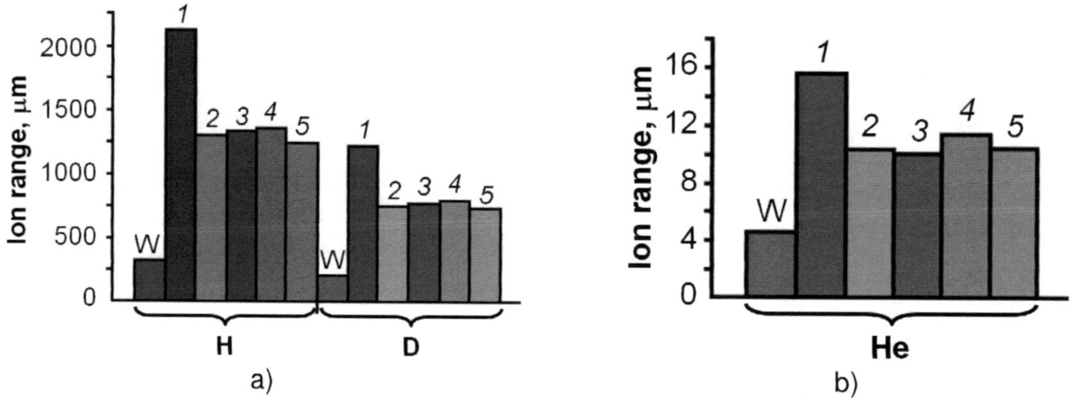

Figure 2. Difference in (a) H^+ and D^+ ions ranges (b) and He^+ ions range in tungsten and in coatings 1–5.

Similar patterns are observed after irradiation by D^+ and He^+ ions also. The type of ion, composition and thickness of the coating mainly affect only the depth of concentration of ions implanted in tungsten.

Capture of hydrogen isotopes and helium of tungsten ions directly depends on the mean free path of the ion in the coating. The difference in the mean free path for hydrogen isotopes helium ions is shown in Figure 2. For all ions, the highest range is observed in coating with the lowest boron content (coating 1). In compounds B-C and B-Ti, the mean free paths are of the close order. It follows from the above that the amount of boron contained in the coating affects the mean free path of implantable ions. An increase in the amount of boron leads to a decrease in mileage in boron-carbon and boron-titanium coatings. In boron-titanium coatings, the range of ions is less than in boron-carbon coatings. The presence of hydrogen in the coating increases the range of ions.

Ions with a lower mass penetrate to a greater depth. The heaviest helium ions have an order of magnitude lower mean free path in coatings than hydrogen isotopes. The penetration depth of ions D^+ is less than that of H^+ ions by ~ 33%.

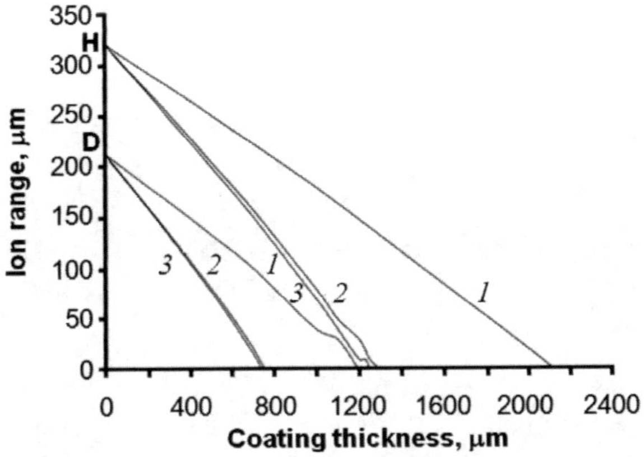

Figure 3. The effect of the composition and coatings thickness on the capture of H^+ and D^+ ions by tungsten.

Titanium boride coatings affect the capture of hydrogen isotopes by tungsten to an equal extent with boron-carbon coatings (Figure 3). Complete absorption of H ions by a coating is possible at a thickness of 1200 μm, and D^+ ions at a thickness of 800 μm. Compound B-C:H requires a thickness of 1.5 times greater for the complete absorption of implantable D^+ ions and 1.8 times for H^+ ions.

4. Conclusion

Studies have shown that titanium boride coatings have an effect on the capture of hydrogen and helium isotopes ions similar to boron-carbon coatings. Given the high stability of titanium boride to cathodic atomization, its thermal and corrosion resistance, such coatings can be used to protect plasma-facing materials.

Acknowledgments

This work was supported by the grant of the Ministry of Education of the Republic of Kazakhstan No. AP05134758 and by Competitiveness Growth Program of the Federal Autonomous Educational Institution of Higher Education National Research Nuclear University MEPhI (Moscow Engineering Physics Institute).

References

[1] McCracken G M and Stott P E 1979 *Nucl. Fusion* **19(7)** 889
[2] Cohen S, Cecchi J, Daughney C et al. 1982 *J. Vac. Sci. Thechnol.* **20(4)** 1226–29
[3] Causey R A 2002 *J. Nuc. Mater.* **300** 91
[4] Liasai S, Funahashi A, Nagami M and Sugie T 1979 *Nucl. Fusion* **19(2)** 195
[5] Kislitsin S, Uglov V and Larionov A 2016 *Proc. 21st Int. Symp. on Heavy Ion Inertial Fusion* (Astana, Republic of Kazakhstan) p 49
[6] Kislitsin S, Uglov V and Larionov A 2016 *AIP Conference Proceedings* **1783** 020091
[7] Kislitsin S, Larionov A, Ivanov I and Zdorovets M 2018 *Proc. of XII Int. Conf. ION-2018* p 41
[8] Gavarini S, Toulhoat N, Peaucelle C, Martin P, Mende J, Pipon Y and Jaffrezic H 2007 *J. Nucl.Mater.* **362** 364
[9] Meshcheryakov A I and Vafin I Yu 2015 *Applied Physics* **3** 15 (in Russian)
[10] Moshkunov K A, Kurnaev V A, Rusinov A A et al. 2009 *Problems of Atomic Science and Technology, Series Thermonuclear Fusion* **4** 24 (in Russian)
[11] Karpikov A N ., Kislitsin S B, Larionov A S and Chekushina L V 2019 *Proc. of the XVI Int. Scientific School-Conf. on Fundamental and Applied Materials Science* p 200 (in Russian)
[12] Larionov A S, Zhakanbaeyev E A, Kozlovski A L, Dikov A S, Kislitsin S B and Chekushina L V 2018 *Eurasian J. of Physics and Functional Materials* **3 (1)** 251
[13] Buzhinskij O and Semenets Yu 1997 *Fusion Technology* **32** 1
[14] Buzhinsky O I, Azizov E A, Belov A M et al. 1992 *J. Nucl. Mater.* **191–194** 1413–16

IOP Conf. Series: Materials Science and Engineering 1005 (2020) 012006 doi:10.1088/1757-899X/1005/1/012006

Oxidation of boron-titanium thin film coating during cyclic tests on thermal shock

E E Suslov[1], A S Larionov[1], S B Kislitsin[1,2], I I Chernov[2], M S Staltsov[2], A S Dikov[1] and V V Firsova[1]

[1] Institute of Nuclear Physics of the Republic of Kazakhstan, Ibragimov st. 1, 050032 Almaty, The Republic of Kazakhstan

[2] National Research Nuclear University MEPhI (Moscow Engineering Physics Institute), Kashirskoe highway 31, 115409 Moscow, Russia

E-mail: larionov-inp@yandex.kz, i_chernov@mail.ru

Abstract. The results of experiments on cyclic thermal shock tests of promising neutron-absorbing titanium boride coatings are presented. It has been established that the main effect is the high-temperature oxidation of titanium boride with the formation of titanium and boron oxides. The regularities of changes in the oxidation parameters during cyclic thermal shock tests are determined.

1. Introduction

Coatings from titanium boride Ti_2B_5 are intended for use as neutron-absorbing coatings on container dtructural materials for moving and densified safekeeping of spent nuclear fuel [1–3]. These coatings can significantly increase the efficiency of attenuation of the neutron flux with a decrease in thickness in comparison with existing analogues [4–6]. Coatings in this case should have a sufficient margin of safety, corrosion and thermal resistance.

The most studied and close to Ti_2B_5 modification of titanium boride is titanium diboride TiB_2 [7]. It is known that titanium diboride has high hardness and chemical resistance, as well as a high melting point of 3225 ± 25 °C. Compounds of titanium boride with a high content of boron such as Ti_2B_5 are less detail studied. This is also confirmed by the state diagram of the B–Ti system [8]. Titanium diboride coatings are used in microelectronics and in the manufacture of cutting tools. As a coating material for neutron absorption deposited by the magnetron method, such coatings have not previously been considered. In this regard, the properties of such films were first studied in this work.

It is known that Ti_2B_5, like TiB_2, has high corrosion resistance since an oxide film (borate) is formed on the surface of these compounds [9, 10]. Titanium diboride is prone to high temperature oxidation with the formation of oxide phases of borate B_2O_3 and rutile TiO_2, as well as titanium borate $TiBO_3$ [11–13].

This paper presents an analysis of the evolution of the structure and phase composition of boride-titanium coatings with a high content of boron.

2. Materials and methods

Austenitic stainless steel 12Cr18Ni9Ti was used as a coating substrate. The samples were cut by spark cutting, the surface was subjected to mechanical grinding and polishing. The size of the samples for thermal shock tests is 20×5×1 mm^3, and the surface roughness is not less than R_a = 80 nm. Magnetron sputtering of coatings was carried out according to the proven technique from a composite boron – titanium target, which is made by hot pressing of titanium and boron carbide powders in a weight ratio of 1:1 on a VUP-5M magnetron device [6]. The power supplied to the magnetron was about 42 W and the sputtering time was 1 hour.

According to X-ray diffractometry, the coating represents the boron – titanium compound in the form of the ω-phase of titanium hemipentaboride Ti_2B_5 with a small amount of impurity β-phase of titanium dodecaboride TiB_{12}. Integral porosity averaged ~ 1%. It contains impurities of carbon and oxygen (~ 26%). Oxygen is located in defects (pores) mainly, as well as in the form of a thin (not more than 10 nm) oxide film (borate). The phase composition of the oxide film by the applied methods could not be determined due to its small thickness; the intensity of X-ray reflexes is too low for analysis. The coating thickness on the samples for study was ~ 200 nm.

Annealing was carried out in the temperature range of 300–900 °C in the SNOL 7.2 / 1100 muffle furnace in an air medium. The coated sample was placed in a furnace heated to a certain temperature for thermal shock tests. The heating rate of the sample under such conditions was 20–60 s. Cooling occurred with the furnace, the cooling time from 900 to 20 °C was ~ 6 h. The temperature of the sample was controlled using a chromel-alumel thermocouple.

The change in elemental composition after testing was monitored by energy dispersive analysis (EDA) using an EDA attachment on a HITACHI TM3030 scanning electron microscope (SEM). X-ray phase analysis (XRD) was carried out on a D8 ADVANCE ECO diffractometer (Bruker, Germany) using CuK_α copper tube radiation. To identify the phases and study the crystal structure, BrukerAXSDIFFRAC.EVAv.4.2 software and the ICDD PDF-2 international database were used. The conditions for obtaining of X-ray diffraction patterns were as follows: Voltage is 40 kV, Current is 25 mA, 2θ = 20–90°, step is 0.03°, standing time at the point is 1 s, a nickel soller was used as an absorber. The depth of X-rays penetration into the material is 8–10 μm under these shooting conditions. The surface morphology of coating was studied by SEM JEOL JSM7500F.

3. Results and discussion

The coating consists of particles with diameters of 100–200 nm and it can be seen that it repeats the relief of the substrate (Figure 1a).

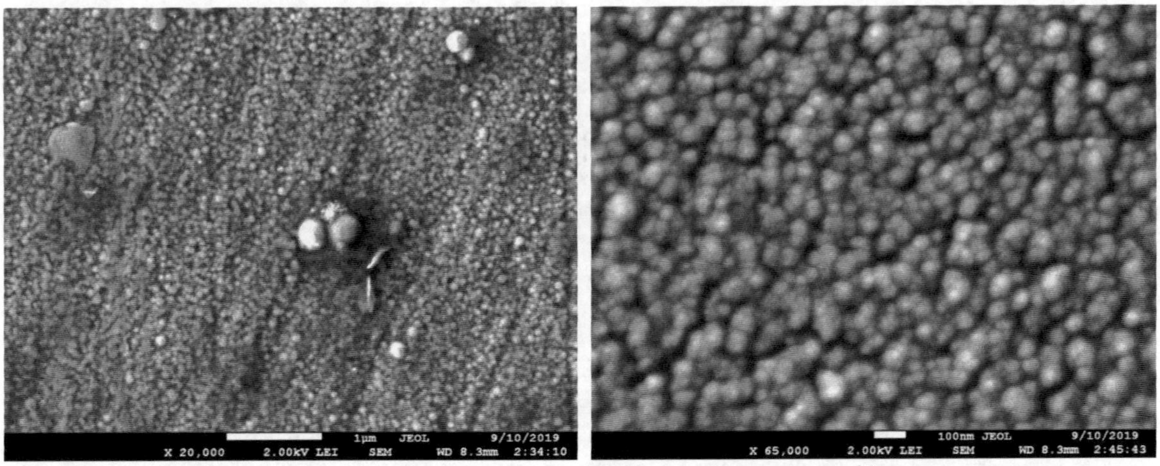

a) b)

Figure 1. Pictures of coating surface at different scales.

During testing of samples with a titanium boride coating for thermal shock, it turned out that the main effect of temperature exposure in air is oxygen saturation of the coating and the formation of scale on the surface. The dependences of these effects on temperature and the duration of its exposure were determined.

In order to reveal the temperature dependence of the manifestation of effects, annealing was carried out to temperatures of 300, 500, 700 and 900 °C. The maximum temperature was chosen taking into account the fact that the recrystallization of 12Cr18Ni9Ti austenitic steel, which is the substrate material, occurs at a temperature above 900 °C [14]. Since recrystallization is one of the determining factors in the operation of structural materials, a further increase in temperature is not a priority from the point of view of practical use. Scale formation was observed after heating at 900 °C only. An increase in the number of cycles led to the formation of cracks and exfoliation of the oxide layer (flaking). An increase in the oxygen content was observed until the 3rd test cycle, then the oxygen content decreased, which was caused by the destruction of the oxide layer (Table 1). In the case when no scale was formed, violations of the structure of the titanium boride coating were not observed.

Table 1. Oxygen content in the titanium boride coating after cyclic tests for thermal shock at 900 °C.

The number of cycles	Before the test	1	2	3
The oxygen content, at. %	26.8	47.3	61.5	47.9

Cyclical tests for thermal shock at temperatures of 300, 500 and 700 °C did not lead to a change in the phase composition of titanium boride coatings. The phase composition changes after the first test cycle at 900 °C. The position of the reflections in the diffraction patterns indicates the presence of phases corresponding to the TiO_2 and B_2O_3 oxides in the (101) and (022) planes, respectively. As the cycles increase, redistribution of the intensities of diffraction reflections of the main identifiable phases occurs (Figure 2).

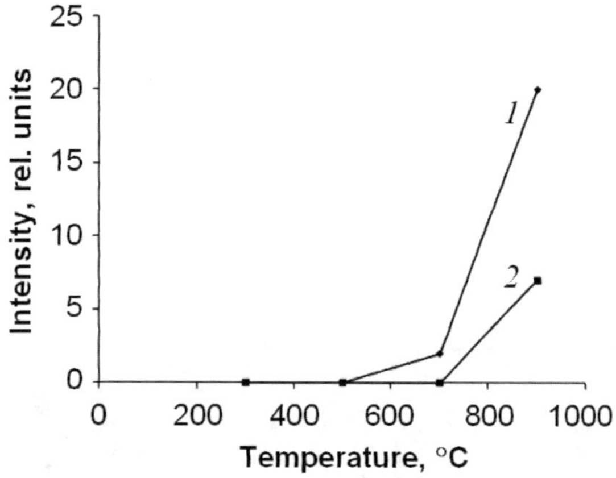

a)

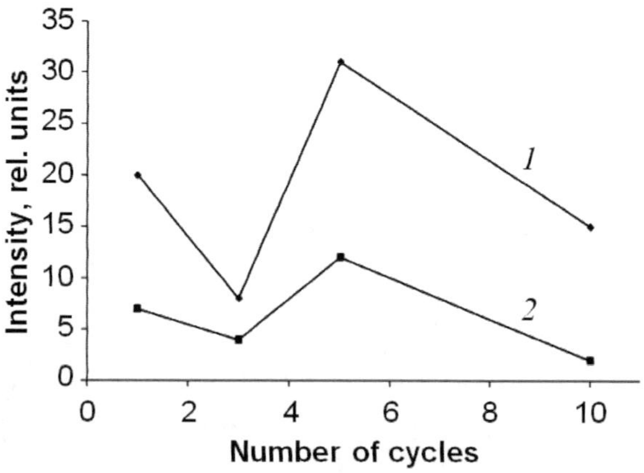

b)

Figure 2. Oxide phases reflections intensity change in dependence on (a) temperature and (b) number of cycles: *1* – rutile; *2* – borate.

All diffraction patterns contain reflections corresponding to the austenitic phase, which gives the material of the steel substrate. Titanium oxide is represented in the hexagonal modification of rutile. Iron oxide (hematite) appears during the oxidation of iron in a stainless steel substrate. Boron oxide (borate) has an amorphous structure.

Borate has a melting point of 320–450 °C. The increase in test cycles leads to its burnout. A change in the intensity of the B_2O_3 phase reflexes shows that the borate formed during the first cycle disappears practically after three cycles. After 5 cycles, the boron oxide reflex appears again, but its

IOP Conf. Series: Materials Science and Engineering 1005 (2020) 012006 doi:10.1088/1757-899X/1005/1/012006

intensity decreases after 10 cycles. The intensity of the titanium oxide reflex varies in the same proportion (see Figure 2b).

After 1 cycle, the scale is not evenly distributed on the surface of the titanium boride coating. After 3 cycles, the scale covers the entire surface, areas appear on which exfoliation occurs. After 5 and 10 cycles, a new oxide layer appears in areas with exfoliated scale; as a result, the surface becomes uneven. In Fig. 3 shows the structure of the oxide layer after 1 test cycle at different magnifications. It can be seen that already after the first cycle a local exfoliation of the formed oxide layer occurs. The oxide layer has a grain structure with grain sizes from 20 to 500 nm.

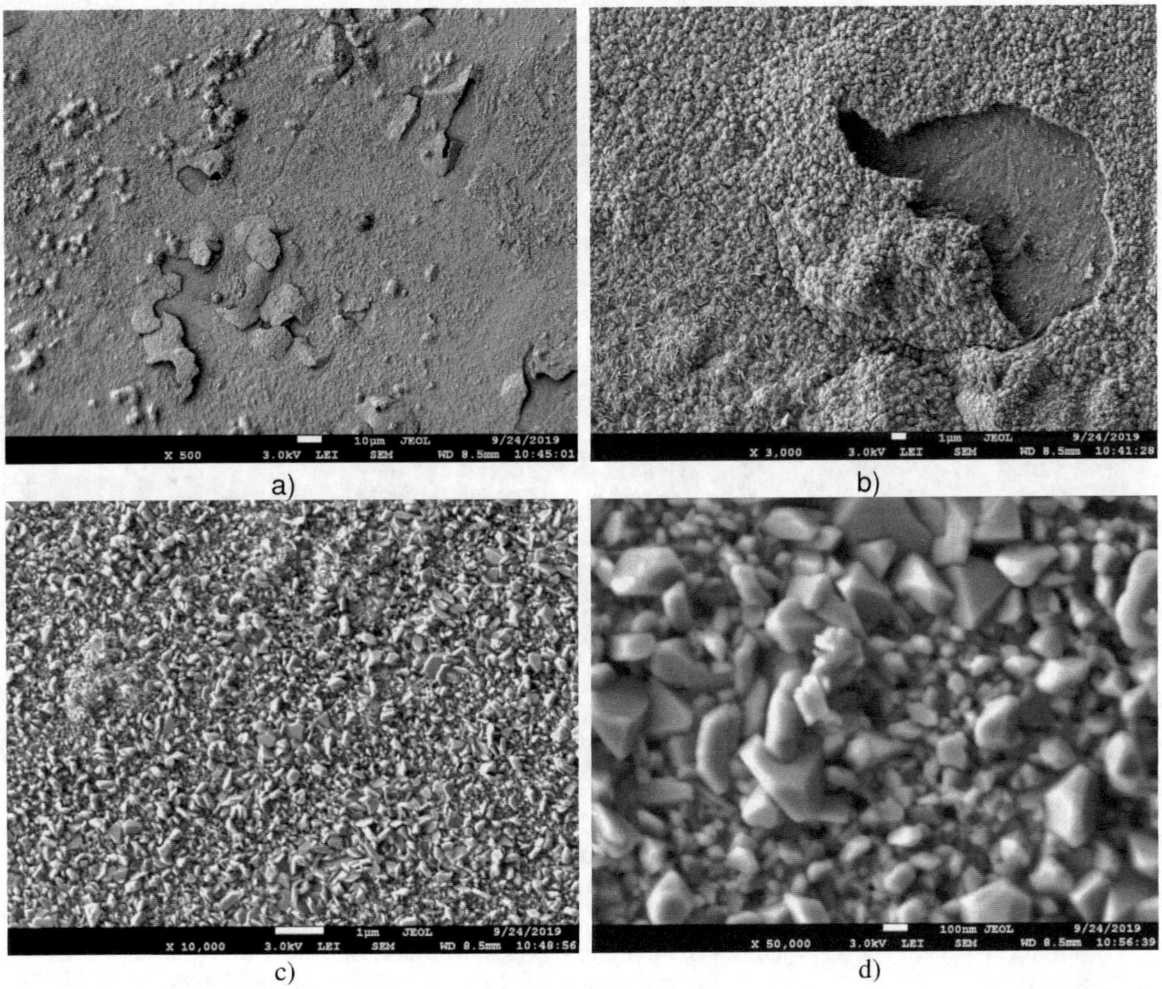

Figure 3. Obtained at different magnifications SEM images on a titanium boride coating oxide film after 1 cycle of thermal shock tests at temperature of 900 °C.

After 3 test cycles, the structure of the oxide film has not fundamental differences. Part of the film is exfoliated and areas without an oxide layer are visible in Figure 4. The grain structure of the oxide film does not change. Figure 5b shows a surface area with visible grains of titanium boride after 5 test cycles. It can be seen that the effect of temperature led to a change in the shape of the grains and a decrease in their average size by ~ 1.5 times. These changes have led to compaction of the grain structure of titanium boride.

IOP Conf. Series: Materials Science and Engineering 1005 (2020) 012006 doi:10.1088/1757-899X/1005/1/012006

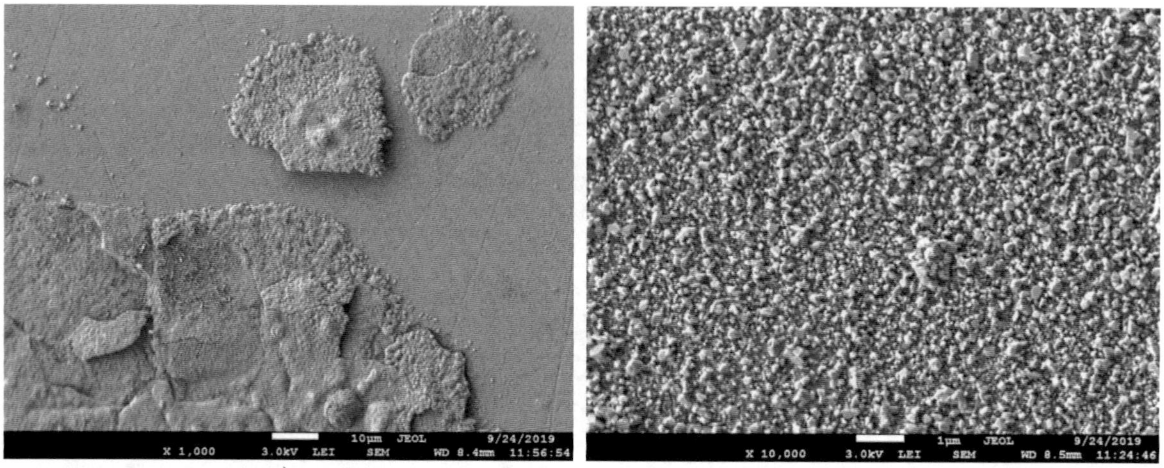

a) b)

Figure 4. SEM images of an oxide film on titanium boride coating after 3 cycles of thermal shock tests at temperature of 900 °C.

a) b)

Figure 5. SEM images of an oxide film on titanium boride coating after 5 cycles of thermal shock tests at temperature of 900 °C: (a) scale view and (b) titanium boride grains.

From the above results we can understand the following. The mechanism of oxide phase formation occurs through the transport of oxygen atoms deep into the material. The accumulation of oxygen at the boundary of the oxide film and titanium boride during heating causes severe embrittlement of the boundary. This mechanism causes flaking and the gradual destruction of the coating. Oxidation in the presence of a large amount of oxygen can occur by reaction

$$2Ti_2B_5 + 23O \rightarrow 5B_2O_3 + 4TiO_2.$$

4. Conclusion

Based on the results, we can draw the following conclusions.

Titanium boride coating is prone to high-temperature oxidation in air at temperature of 900 °C as showed investigations of its thermal stability by cyclic thermal shock tests.

Formation of the oxide phase and destruction of titanium boride coating were not detected at temperatures below 700 °C.

Since the storage of spent nuclear fuel involves a temperature of about 350 °C, the characteristics of the developed coating make it possible to use it under these conditions.

Acknowledgments
This work was supported by the grant of the Ministry of Education of the Republic of Kazakhstan No. AP05134758 and by Competitiveness Growth Program of the Federal Autonomous Educational Institution of Higher Education National Research Nuclear University MEPhI (Moscow Engineering Physics Institute).

References

[1] Dvilis E S, Khasanov O L, Gulbin V N, Petyukevich M S, Khasanov A O and Olevsky E A 2016 J. *of Minerals. Metals & Materials Society* **68(3)** 908
[2] Choi J, Lee C, Day D, Wall M, Saw C, MoberlyChan W, Farmer J, Boussoufl M, Liu B, Egbert, Branagan D and D'Amato A 2015 *Applications of Neutron-Absorbing Structural-Amorphous Metal (SAM) Coatings for Criticality Safety Controls of Used Fuel Storage, Transportation, and Disposal* (ICNC 2015, Charlotte, NC, United States) 23 pp
[3] Mizia R E, Wright R N, Swank W D, Lister T E and Pinhero P J 2007 *United States Patent* No. US 7,286,626 B2
[4] Larionov A S, Zhakanbaeyev E A, Kozlovski A L, Dikov A S, Kislitsin S B and Chekushina L V 2018 *Eurasian J. of Physics and Functional Materials* **3(2)** 251
[5] Larionov A S, Chekushina L V and Suslov E E 2019 *Materials Science Forum* **945** 660
[6] Akayev S O, Larionov A S and Karpikov A N 2019 *Bulletin of KazNaTU* **1** 229 (in Russian)
[7] Samsonov G V, Serebryakova T I and Neronov V A 1975 *Borides* (Moscow: Atomizdat) 376 pp (in Russian)
[8] Computational Thermodynamics CALPHAD (Calculation of Phase Diagrams) [electronic resource] Date of access: 09.02.2015 Mode access: http://www.calphad.com/tinium-boron.html
[9] Li Q., Lai Y, Li J, Yang J, Fang J and Chen Z 2005 *Light Metals* **2005** 789
[10] Kontrík M, Šimko F, Galusková D, Nosko M, Bizovská V, Hičák M, Galusek D, Rakhmatullin A and Korenko M 2018 *J. of European Ceramic Society* **38(4)** 1143–51
[11] Hurlen T J 1960–1961 *J. Institute of Metals* **89 (4)** 128
[12] Jenkins A E 1955–1956 *J. Institute of Metals* **84 (10)** 1
[13] Sood D K, Mukherjee S, Katselis G, Brown I G, Prince K E, Short K T and Evans P J 1998 *Surface and Coatings Technology* **103&104** 304
[14] Viktorov N A 2011 *Metallurgy and heat treatment of metals* **6** 8 (in Russian)

IOP Conf. Series: Materials Science and Engineering 1005 (2020) 012007 doi:10.1088/1757-899X/1005/1/012007

The effect of neutron irradiation in a helium atmosphere on the mechanical properties of 18Cr10NiTi austenitic steel

S O Akaev[1,2], A S Dikov[1,3], L A Dikova[1], S B Kislitsin[1,3], V V Firsova[1], A S Larionov[1], I I Chernov[3] and M S Staltsov[3]

[1] Institute of Nuclear Physics of the Republic of Kazakhstan, Ibragimov st. 1, 050032 Almaty, The Republic of Kazakhstan
[2] Satbayev University, Satpaev st. 22a, 050013 Almaty, The Republic of Kazakhstan
[3] National Research Nuclear University MEPhI (Moscow Engineering Physics Institute), Kashirskoe highway 31, 115409 Moscow, Russia

E-mail: lexa_edji@mail.ru, i_chernov@mail.ru

Abstract. The microstructure and mechanical properties of 18Cr10NiTi austenitic steel were studied after high-temperature neutron irradiation of samples in a helium atmosphere. It was revealed during irradiation, helium penetrates into steel samples forming bubbles distributed on dislocations and along grain boundaries. Character of the temperature dependence of the mechanical characteristics of irradiated steel indicates a high-temperature radiation embrittlement of steel. The character of the mechanical properties temperature dependence change of irradiated steel points to its high-temperature radiation embrittlement.

1. Introduction

It is known that the presence of helium in structural materials is one of the factors limiting their working capacity in nuclear reactor core [1–3]. One of such manifestations may be the formation of bubbles chains along grain boundaries. This leads to weakening of intergranular bonding and grain boundary cracking known as the phenomenon of high-temperature radiation embrittlement (HTRE) [1, 4, 5]. Along with this, there is a point of view according to which the effect of helium on the mechanism of HTRE is secondary, and it connects HTRE with the segregation of impurity atoms, hardening of the material inside the grains themselves as a result of gas bubbles formation, presence of vacancy clusters, particles of the secondary phases, as well as, possibly, by braking of polygonization and recrystallization processes. As a result, the imbalance in the strength of the grain body and its boundaries increases, leading to grain-boundary destruction of a irradiated material [6–8]. These authors justify the denial of predominant role of helium in a metals high-temperature ductility decrease by the fact that HTRE is most pronounced in materials that are able to loss of high-temperature ductility as a result of various treatments even without irradiation [6, 8]. For example [6], in the experiments of helium saturation by a defect-free method using the “tritium trick” method, the role of helium was not established, and the authors considered that HTRE is associated with a impurities state change in the metal and formation of precipitates under irradiation. However, the fact of the effect of helium on high-temperature embrittlement of materials was revealed unambiguously using a similar technique for introducing of helium into 304L and 309S steels, vanadium based alloy, niobium and its alloys and titanium alloys [5]. Despite of different views on the HTRE mechanisms, most researchers attribute it to the influence of helium atoms formed in the reactor during irradiation [1–3, 5, 9–12]. It was found in these works the effect of helium on HTRE is significant, and helium is determining in some cases, and the main mechanism of this effect is the accumulation of helium at elevated temperatures at the boundaries in the form of bubbles chain.

Thus, it is not always easy to differentiate the effect of helium and other processes on the mechanism of HTRE. This is due to the fact that the helium accumulation rate correlates with the rate of radiation defects generation in steels, which, in turn, affect radiation-stimulated processes, including diffusion and segregation phenomena [13, 14]. It is known only for certain that the HTRE phenomenon manifests itself only in polycrystallines at temperatures above 0.45 of melting temperature [2, 5].

Ion implantation methods and a tritium trick are widely used in addition to high-dose neutron irradiation to study the effect of helium on the properties of various materials [2, 5, 15, 16]. However, these methods do not allow to unambiguously isolate the role of helium in HTRE also. In addition, when using various methods to simulation of the HTRE process, there are a number of problems: high radioactivity of the irradiated material (neutron irradiation); small depth of helium penetration (ion-implantation of helium); generation of a large number of radiation defects (simulating irradiation by heavy ions with the simultaneous introduction of helium ions); hydrogenation, which also leads to embrittlement of the material (tritium trick).

Therefore, it is of interest to consider a helium accumulation in structural materials under high-temperature neutron irradiation during high-temperature neutron irradiation in a helium atmosphere up to damaging dose of ~ 1 dpa, i.e. at relatively low concentrations of introduced radiation defects.

2. Experimental procedure

The material for research was the austenitic steel 18Cr10NiTi in the state of 10% cold-worked (CW). This type of steel is widely used in nuclear reactors as a structural material for nuclear reactor internals. Investigation of helium penetration into steel during high-temperature neutron irradiation was carried out using gas-filled samples (ampoules). The ampoule wall thickness was 2 mm; the helium pressure during irradiation did not exceed atmospheric pressure. The irradiation temperature was 1033 K. Warming-up of ampoules was carried by neutron irradiation only. There was no contact of the ampoules with coolant during irradiation. Neutron irradiation was performed in the WWR-K reactor up to fluence of $9 \cdot 10^{23}$ n/m^2 at neutron flux (E > 0.1 MeV) of $7.6 \cdot 10^{16}$ n/(m$^2 \cdot$s) [17]. Samples for research were prepared from the walls of irradiated ampoule. The samples cutting scheme is shown in Figure 1.

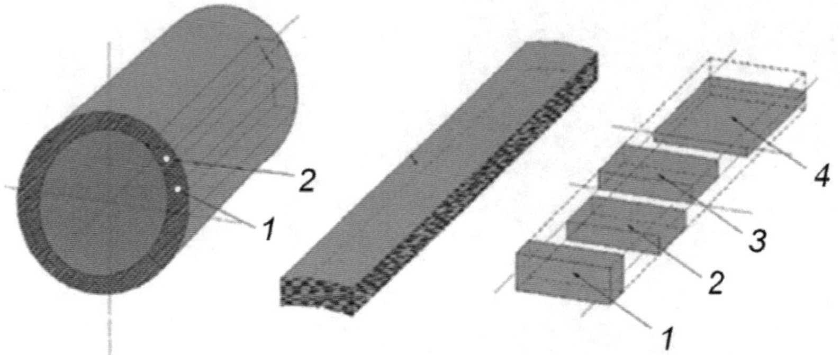

Figure 1. Scheme of samples cutting for research: *1–3* are samples for thermal desorption spectrometry, *4* are samples for electron microscopy.

To eliminate of cold-hardening caused by mechanical action, the samples were subjected to the standard preparation procedure by mechanical grinding and polishing and final electro polishing in a solution of 20% HClO4 + 80% C2H5OH [18]. Samples for TEM were prepared by jet electrolytic thinning in the same acid solution.

The presence of helium in steel after neutron irradiation and the peculiarities of its release were determined by helium thermal desorption spectrometry (HTDS) during uniform by rate heating of 0.7 K/s in the temperature range of 300–1273 K.

The microstructure of the irradiated samples was studied in TEM JEOL JEM 2100.

Mechanical characteristics of irradiated samples were determined by tensile tests using the LR5K Plus universal testing machine at temperatures of 298, 723 and 1033 K. After mechanical tests, the nature of fracture was studied in Hitachi TM4000Plus scanning electron microscope

3. Results and Discussion

Figure 2 shows the HTDS spectra irradiated steel. TEM image of steel after irradiation is presented on Figure 3.

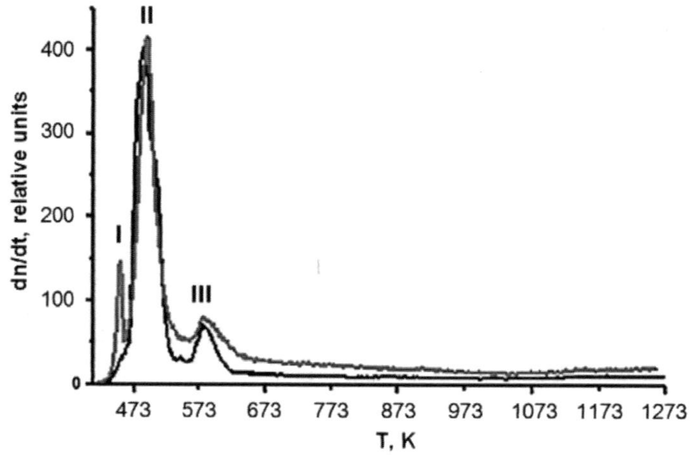

Figure 2. HTDS spectra of 18Cr10NiTi steel irradiated by neutrons in a helium atmosphere at a temperature of 1033 K.

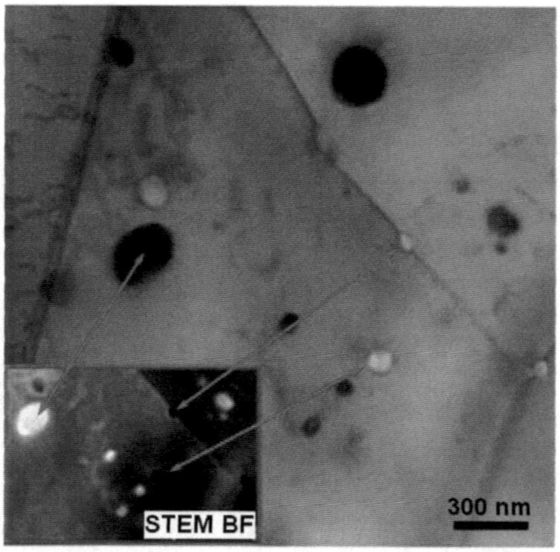

Figure 3. Microstructure of 18Cr10NiTi steel irradiated by neutrons in a helium atmosphere at a temperature of 1033 K.

As can be seen in the HTDS spectra (see Figure 2), there are three peaks of helium release of different intensities in the temperature range of 300–625 K, which indicates the existence of several discrete stages with differing mechanisms of gas release during heating [19, 20]. Helium is insoluble in metals [5], it has extremely high mobility (for example, the migration activation energy of helium atoms is only 0.08 eV in Ni [22]), therefore, it may be trapped and retained with an sufficiently high binding energy on imperfections in the crystal structure only (vacancies, pores, interfaces between the second phases and the matrix, dislocations, grain boundaries and twins, etc.). The low-temperature location of HTDS peaks indicates a very low binding energy of introduced helium atoms with those of traps on which they are remain in irradiated steel up to temperature of ≤ 625 K.

Table 1 shows the activation energies of helium release in the TDS peaks in Figure 2, calculated by the Reherd formula [23].

Table 1. Gas release activation energies in the HTDS peaks

	Peak number		
	I	II	III
E^a, eV	2.1 ± 0.2	2.2 ± 0.2	2.5 ± 0.3

Accompanying all HTDS spectra first peak is observed at temperature of 453 K with low activation energy of helium desorption of 2.1 eV only. It is most likely due to the release of helium concentrated at a very shallow depth near the surface or adsorbed by the sample surface, since the binding energy of helium atoms with other possible defects is much more higher in metals [24, 25]. It should be noted that this peak does not always manifests itself clearly on different spectras of HTDS (see Figure 2).

The second and third peaks were detected at temperatures of 503 and 593 K respectively. Peak temperatures are too low compared to the irradiation temperature (T_{irrad} = 1033 K). Therefore, it is difficult to identify their nature, but some assumptions can be made based on a number of known facts.

1. Complexes of a He_mV_n type are formed (m is the number of He atoms, n is the number of vacancies in the complex) during low-temperature (room temperature) introduction of helium. This complexes are thermally stable in nickel up to ~ 923 K at m ≈ n [24, 25].The possibility formation of thermally more stable complicated complexes of the $He_mM_kV_n$ type (k is the number of atoms of the alloying element in the complex) with a dissociation temperature of about 1033 K is shown after alloying of nickel by "oversized" elements (M = Al or Ti) [26]. However, T_{irrad} is 1033 K in the present experiment; therefore, the formation of any helium-vacancy complexes is impossible, and helium bubbles (gas-filled pores) form immediately during irradiation by two- or polyatomic nucleation of their [5].

2. Having low solubility and high mobility, introduced helium "searches" traps for yourself, the most probable of which can be dislocations (steel was in the stat of 10% CW) and grain boundaries in the homogeneous 18Cr10NiTi steel (particles concentration of primary TiC carbide is negligible [2]). As can be seen on Figure 3, this is confirmed by the presence of gas pores (the dose of 1 dpa is too small for void formation [2, 27]) along the boundaries of austenitic grains, at triple junction points of grains and in the places of structural defects (dislocations) accumulation: gas-filled pores / bubbles appear light on the bright-field TEM image and they look as dark objects with a light border on the dark-field STEM image (shown in the insert).

Thus, the appearance of peaks II and III can be attributed to release of helium along grain boundaries and along dislocations [5, 19, 28]. Intensity of helium desorption in peak II significantly exceeds the intensity of helium release at stages I and III. Probably, this is associated with the gas outlet from the volume to surface along grain boundaries in peak II [5, 19]. But clarification required in what form helium releases grain boundaries (atomic, in the form of bubbles, or something else).

IOP Conf. Series: Materials Science and Engineering 1005 (2020) 012007 doi:10.1088/1757-899X/1005/1/012007

The manifestation of the peak of helium desorption at a temperature of 593 K remains to be associated with the release of helium through dislocation tubes [5, 19, 28, 29].

Figure 4 presents the results of mechanical uniaxial tensile tests of 18Cr10NiTi steel irradiated by neutrons in helium atmosphere. Figure 4 presents the results of mechanical uniaxial tensile tests of 18Cr10NiTi steel irradiated by neutrons in helium atmosphere. As can be seen, decrease in the ultimate tensile strength (σ_{UTS}) and relative elongation (δ) are observed with increasing of test temperature T_{test}. Compared to investigation at room temperature, σ_{UTS} decreases by ~ 16% and value of δ by ~ 19% at T_{test} = 723 K. A further increase in T_{test} to 1033 K leads to softening of steel by ~ 64%, and a decrease in ductility by 53%. However, the change in the yield strength (σ_Y) as a function of T_{test} is not monotonic: σ_Y of steel increases by 25% with an increase in T_{test} from 298 to 723 K. On the contrary, the test at T_{test} = T_{irrad} = 1033 K reduces it by 56%. As was shown earlier [30], an increase in σ0.2 at T_{test} = 723 K can be associated with structural changes occuring in this irradiated by neutrons steel at a given temperature.

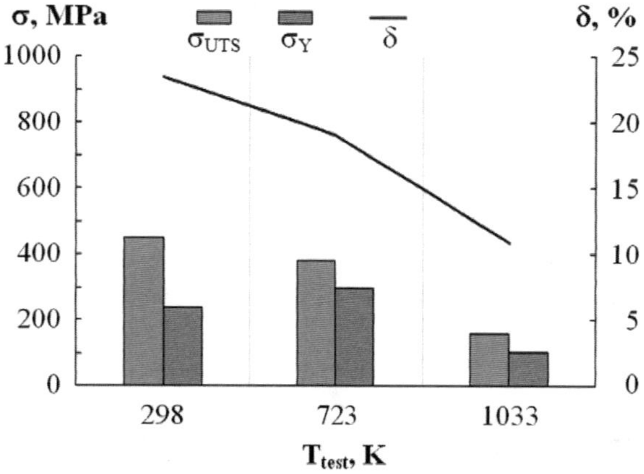

Figure 4. The dependence of mechanical properties on test temperature of 18Cr10NiTi steel irradiated by neutrons in a helium atmosphere.

Such behavior of the mechanical properties of steel saturated by helium in the process of high temperature neutron irradiation may indicate the occurrence of HTRO typical for fcc materials in the presence of helium, and this confirms the role of helium in this phenomenon [1–3, 5, 9–12].

The results of fractographic analysis of steel after mechanical tests are presented in Figure 5. There are a lot of small and large pores forming a dimpled relief in the fracture of the sample tested at room temperature (see Figure 5a). The nature of the fracture is viscous, like a “cup-cone” type. Fracture character is changed to brittle-ductile with increasing test temperature to 723 K (see Figure 5b). Destruction occurs predominantly along the grain boundaries, as evidenced by the presence of "tearing" formed at the site of carbide inclusions located along grain boundaries. With increasing Tisp to 1033 to the fracture surface is a kink with a plurality of flat or slightly concave surfaces), and there are no obvious signs of plastic deformation (see. Figure 5c). There are cracks propagated by a brittle-viscous mechanism with a predominance of a brittle component. It should be noted that such a mechanism characteristic for materials destroyed at low temperatures [31].

And finally, you should mention another aspect relating to the mechanism of HTRE. Steel 18Cr10NiTi was entire period of irradiation in the reactor for 3280 h at T_{irrad} = 1033 K. This temperature is close to the temperature of maximum rate of high-chromium carbide of $M_{23}C_6$ type precipitation in austenitic steel [2]. However, we did not found significant traces of carbide inclusions, except for primary carbides / carbonitrides Ti (C, N), either in TEM studies of the microstructure or in fractographic studies of samples fractures. Titanium is added to steel in an amount of $C_{Ti} \geq 5 \cdot (C_C - 0.03\%)$ (C_C is carbon concentration in percents) in order to prevent intercrystalline corrosion by binding all carbon in TiC carbide so that chrome doesn't get carbon for the formation of $M_{23}C_6$ carbide along the grain boundaries. Thus, the effect of the second phases precipitates on the HTRE including grain boundary phase particles as suggested by the authors of several studies [1–3, 5, 9–12], is excluded and the decrease in ductility during high-temperature tensile tests was due to the presence of helium precisely.

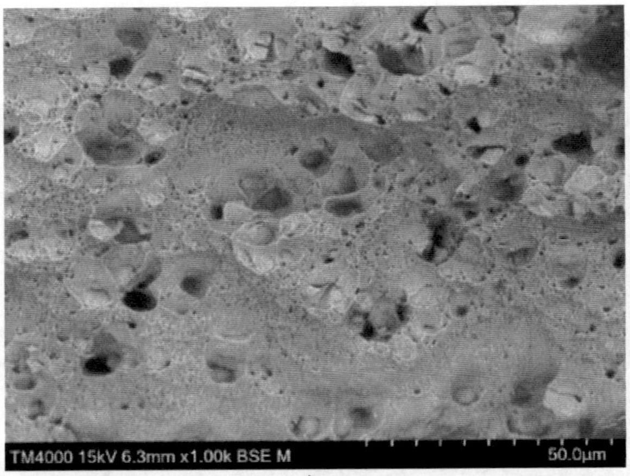

a)

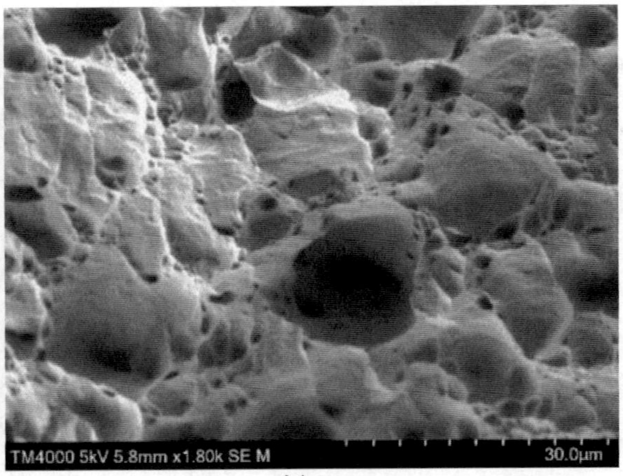

b)

c)

Figure 5. Fractograms of breaking plases of 18Cr10NiTi steel after mechanical testing at temperatures of 298 (a), 723 (b) and 1033 K (c).

4. Conclusions

The following conclusions can be made based on the results of studies influence of helium introduced into 18Cr10NiTi steel by an unusual method of ballistic recoil during the bombardment of a gas-filled balloon by fast neutrons.

1. At high temperature (1033 K) introducing of helium, it is trapped predominantly on structural defects with a low formation energy – grain boundaries and dislocations.
2. Helium release from irradiated steel occurs at very low temperatures with low activation energies during uniform postradiation heatings. This indicates a low binding energy of helium with these traps at high temperature.
3. The dependence of mechanical properties on test temperature of steel irradiated in a helium environment shows the manifestation of a high-temperature decrease in the ductility of steel in the presence of injected helium.

Acknowledgements

This work was carried out as part of the targeted funding program of the Ministry of Education and Science of the Republic of Kazakhstan No. BR05236400 and was supported by Competitiveness Growth Program of the Federal Autonomous Educational Institution of Higher Education National Research Nuclear University MEPhI (Moscow Engineering Physics Institute).

References

[1] NogamI Sh, Hasegawa A, Tanno T, Imasaki K and Abe K 2011 *J. Nucl. Sci. and Technology* **48** 130
[2] Kalin B A, Platonov P A, Tuzov Yu V, Chernov I I and Shtrombah Ya I 2012 *Physical Materials Science* (*Structural Materials for Nuclear Technology* vol 6) ed B A Kalin (Moscow: NRNU MEPhI) 736 pp (in Russian)
[3] Odette G R and Zinkle S J (Eds.) 2019 *Structural Alloys for Nuclear Energy Applications* (Amsterdam: Elsevier) 656 pp
[4] Barnes R S 1965 *Nature* **206** 1307–10
[5] Zaluzhnij A G, Sokurskij Yu N and Tebus V V 1988 *Helium in Reactor Materials* (Moscow: Energoatomizdat) 224 pp (in Russian)
[6] Votinov S N, Prokhorov V I, Balashov V D and Shamardin V K 1970 *Solid State Radiation Physics and Reactor Materials Science* (Moscow: Atomizdat) pp 82-94 (in Russian)
[7] Blackburn L D, Ward A L and Steichen J M 1977 *Proc. Int. Conf. on Radiation Effects in Breeder Reactor Structural Materials* (Scottsdale: Arizona, USA) p 317
[8] Arbuzov V L, Votinov S N and Grigoryan A A 1983 *Atomic Energy* **55** 214
[9] Nikiforov A S, Zakharov A P, Chuev V I and etc. 1982 *Atomic Energy* **53** 3
[10] Schroeder H, Kesternich W. and Ullmaier H. 1985 *Nucl. Eng. and Design/Fusion* **2** 65
[11] Trinkaus H. 1986 *Radiat. Effects* **101**, 91
[12] Trinkaus H. and Ulmaier H. 1994 *J. Nucl. Mater.* **212-215** 303
[13] Stoller R E 1990 *J. Nucl. Mater.* **174** 289
[14] Ivon P (Ed.) 2017 *Structural Materials for Generation IV Nuclear Reactors* (London: Woodhead Publishing) 664 pp
[15] Ryazanov A, Matsui H and Kazaryan A V 1999 *J. Nucl. Mater.* **271&272** 356
[16] Malkov I L, Grishechkin S K, Zlatoustovsky S V, Lebedev B S and Yukhimchuk A A 2008 *Questions of Atomic Science and Technology. Series: Thermonuclear Fusion* **2** 72 (in Russian)
[17] Shaimerdenov A A, Nakipov D A, Arinkin F M, Gizatulin Sh Kh, Chakrov P V and Kenzhin Ye A 2018 *Physics of Atomic Nuclei* **81** 1408–11
[18] Beckert V and Klemm H 1984 *Handbuch der Metallographischen Ätzverfahren* (Leipzig: VEB Deutscher Verlag für Grundstoffindustrie)
[19] Zaluzhny A G and Suvorov A L. 2001 *J. of Technical Physics* **71(2)** 55 (in Russian)
[20] Ruzhytskyi V V, Karpov S A, Kalchenko A S, Кopanets I E, Sungurov B S, Tolstolutskaya G D 2018 *East Eur. J. Phys.* **4** 69
[21] Wilson W D and Johnson R A 1972 *Interatom. Potent. and Simul. Lattice Defects* (N.Y.-London, Plenum Press) 375
[22] Poker D B and Williams J M 1982 *Appl. Phys. Lett.* **40 (9)** 851
[23] Oku D, Yamada T, Hirohata Y et. al. 2007 *J. Nucl. Mater.* **367–370** 864
[24] Reed D J 1977 *Radiat. Effects* **31(3)**, 129
[25] Donnelly S E 1985 *Radiat. Effects* **90(1/2)** 1
[26] Chernov I I, Kalin B A, Kalashnikov A N and Ananin V M. 1999, *J. Nucl. Mater.* **271&272** 333
[27] Zelensky V F, Nekludov I M and Chernyaeva T P 1988 *Radiation Defects and Swelling of Metals* (Kiev: Naukova Dumka) 296 pp (in Russian)
[28] Gong Y, Jin Sh, Zhu Te, Cheng L, Cao X, You Li, Lu G, Guo L and Wang B 2018 *Nucl. Fusion* **58** IOPscience 046011
[29] Gong Y H, Cao X Z, Jin S X et al. 2016 *J. Nucl. Mater.* **482** 93
[30] Dikov A, Kislitsin S and Chernov I 2017 *Defect and Diffusion Forum* **375** 134
[31] Rogers C B and Clayton J Q 1989 *Metallurgical* **23** 1785–90

Measurement of internal friction of Zr–1%Nb alloy after oxidation and nitriding

V Anan'in, B Kalin and A Sumarokova

National Research Nuclear University MEPhI (Moscow Engineering Physics Institute), Kashirskoe highway 31, 115409 Moscow, Russia

E-mail: muzavi36@mail.ru

Abstract. This paper presents results of research of the Zr–1% Nb alloy using internal friction on sheet samples after oxidation due to leakage of air into the measurement facility, as well as after nitriding in an atmosphere of technical nitrogen. The measured temperature dependences of internal friction have been separated into partial maxima with the calculation of the effective activation energy according to the Wert-Marx formula. The purpose of the work is to detect the maxima of the internal friction of interstitial impurities in the alloy. It has been established that the oxygen maximum in successive measurements has activation energy 225, 222, 204 kJ/mol, and the nitrogen maximum has 234–236 kJ/mol. After lengthy high-temperature measurements, a carbon maximum appeared with activation energy 165 kJ/mol. For the grain-boundary maximum activation energy from 197 to 203 kJ/mol was obtained for different samples, while the maximum height monotonously decreased with oxidation. These results for the Zr–1% Nb alloy were obtained for the first time and can be useful in corrosion tests of zirconium alloys.

1. Introduction

The internal friction method using Snoek relaxation is widely used to study the behavior of interstitial impurities in bcc metals and alloys. In pure hcp metals, Snoek relaxation is not observed due to the lattice symmetry. However, if there is a sufficient amount of substitutional impurities in the alloy, asymmetric distortion of the lattice can occur, and when an alternating load is applied, under certain conditions it is possible to observe the Snook relaxation of interstitial atoms near substitutional atoms [1]. From the literature on internal friction in zirconium, the position of the grain boundary maximum (its activation energy is ~ 241, 288, 222 kJ/mol in various studies), the oxygen maximum in zirconium, and in zirconium alloys with a high oxygen content (activation energy 184 and ~ 192–202, as well as 219–207 kJ/mol are known, respectively) and a nitrogen maximum (activation energy ~ 270 and 241 kJ/mol) [2]. The diffusion parameters of interstitial impurities in alpha-zirconium are also known [3], reliable results of which are shown in Table 1.

Table 1. Activation energies of diffusion of interstitial impurities in zirconium and the calculated position of relaxation maxima (Tm) for an oscillation frequency of 2 Hz.

Impurity	H, kJ/mol	D_0, m^2 / s	Tm, °C
C	152	$2 \cdot 10^{-7}$	350
O	204.4	$3.58 \cdot 10^{-4}$	553
N	231.8; 231.0	$10.2 \cdot 10^{5}$; $1.9 \cdot 10^{5}$	660

IOP Conf. Series: Materials Science and Engineering 1005 (2020) 012008 doi:10.1088/1757-899X/1005/1/012008

Thus, a purpose of this work is to detect relaxation maxima of the internal friction of interstitial impurities in the Zr–1% Nb alloy.

2. Materials and Experimental Methods

Measurement of amplitude-independent internal friction by the method of damped oscillations with a frequency of about 2-2.3 Hz was carried out on a MIFI-9 low-frequency relaxometer with a direct bending pendulum in a vacuum of about 0.013 Pa, using the measuring and processing programs described in [4]. The obtained temperature dependences of internal friction were divided into partial relaxation maxima with the calculation of the effective activation energy according to the Wert-Marx formula [5]. The experiment was carried out on samples with a width of 7 mm, a working length of 20-22 mm, a thickness of 0.3 mm, made of a sheet of alloy Zr-1%Nb in the delivery state. All graphs are presented in coordinates of Q^{-1} versus temperature in °C.

3. Results and Discussion

3.1. Measurements after oxidation

A series of measurements of internal friction was performed on a single sample. Figure 1 and 2 shows the temperature dependence of internal friction (TDIF) of the sample in the initial state and its separation into partial maxima.

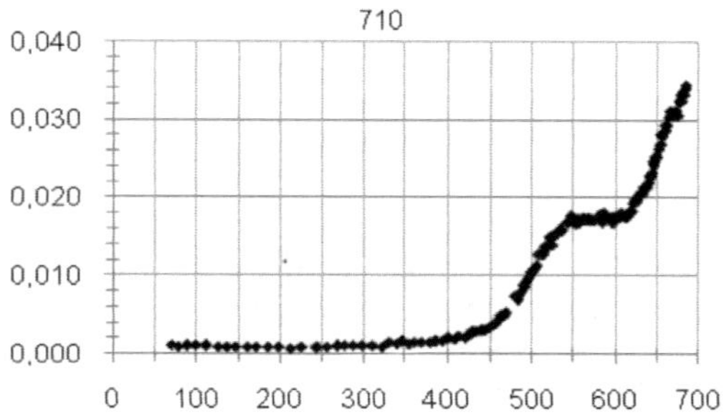

Figure 1. TDIF of the initial sample (No.6).

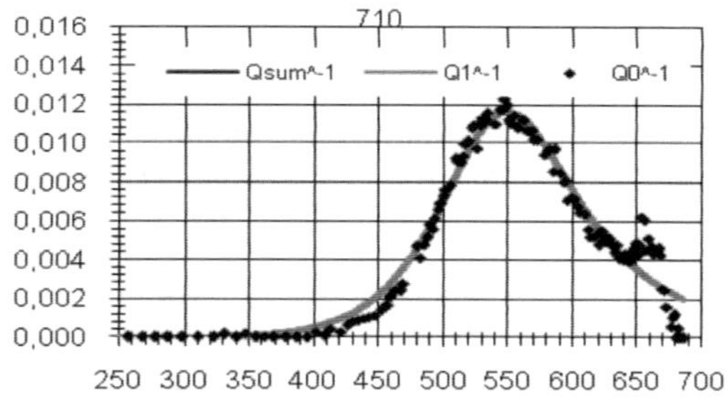

Figure 2. Separation into partial maxima. The grain boundary maximum: Tm – 547 °C, H – 202.7 kJ/mol, $Qmax^{-1}$ – 0.0114.

The results of repeated measurement of sample No.6 (experiments 711 and 712) in terms of separating the curve into partial maxima are shown in Figure 3 and 4.

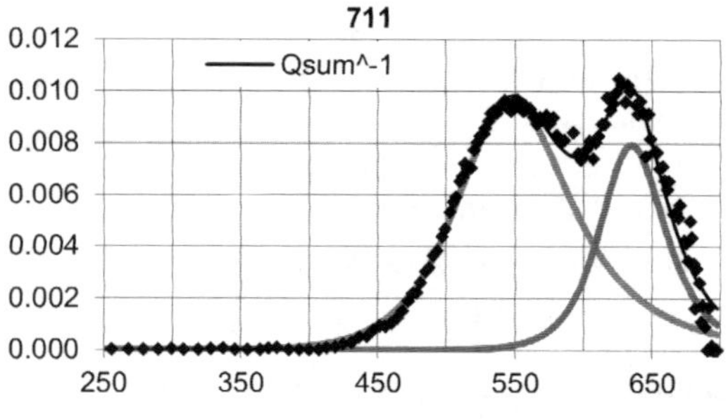

№	T,°C	H, kJ/mol	$Qmax^{-1}$
1	547	202.4	0.00965
2	635	224.8	0.00794

Figure 3. Separation in partial maxima of 711.

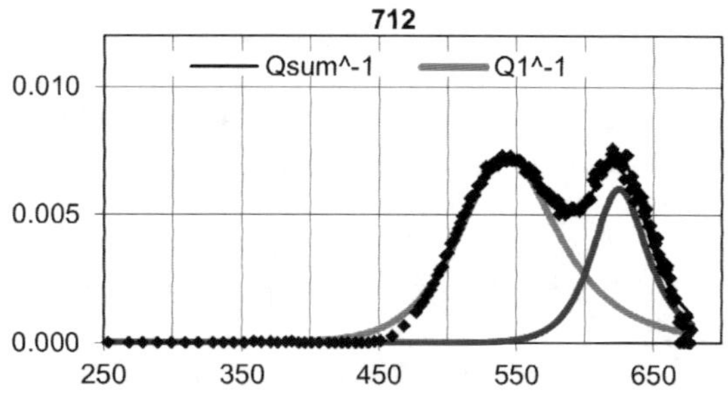

№	T,°C	H, kJ/mol	$Qmax^{-1}$
1	543	201.3	0.00731
2	625	222.3	0.00602

Figure 4. Separation in partial maxima of 712.

A comparative analysis of the data in Figures 1,2 and Figures 3,4 shows that after sufficiently long heating to 700 °C in the previous experiment, in experiments 711 and 712 a maximum appeared, that is close to oxygen in activation energy (see above). The explanation is simple - the sample was oxidized in a bad vacuum during the measurement process. In these two measurements, the position of the maxima and their parameters remained practically unchanged, as did the ratio of the heights of the oxygen maximum to the grain boundary maximum (0.823), with a general small decrease in the height of both maxima.

In the following two measurements, under similar conditions (exp 714 and 715), two more peaks appeared – a low-temperature maximum and a maximum located to the right of the oxygen one. Figure 5 shows the results of experiment 715:

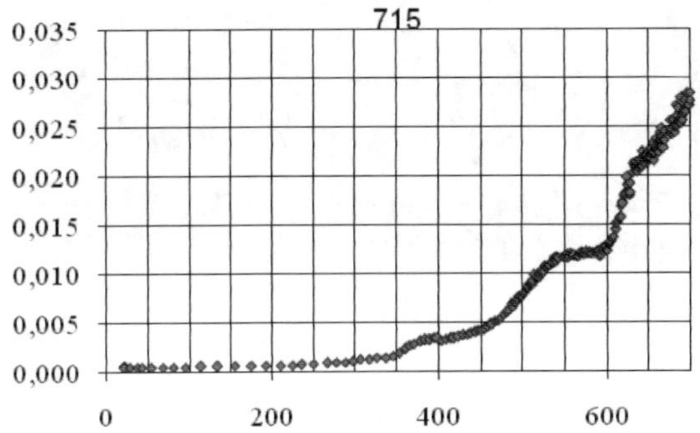

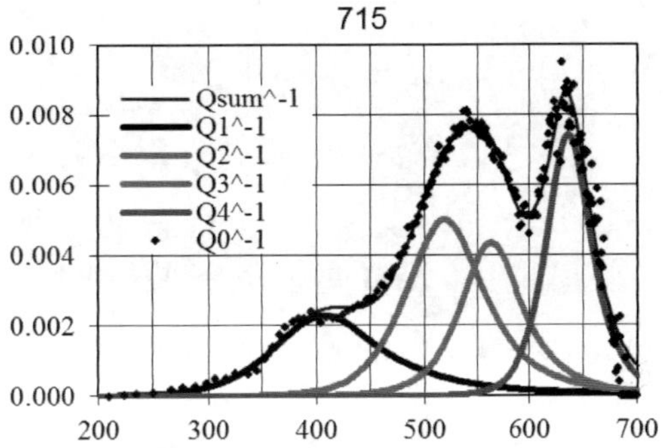

№	Tmax, °C	H, kJ/mol	$Qmax^{-1}$
1	407	166.3	0.00227
2	518	194.5	0.00501

№	Tmax, °C	H, kJ/mol	$Qmax^{-1}$
3	562	205.8	0.00435
4	636	224.6	0.00741

Figure 5. 5th measurement of the sample No.6 in experiment 715 and separation into partial maxima.

Our interpretation of the possible nature of the maxima in experiment 715 is as follows. Maximum No.1, with activation energy 165.4 kJ / mol - in all probability, carbon, because it appeared after repeated high-temperature measurements with steam-oil diffusion pump; it has been observed before. Maximum No. 2 – grain-boundary maximum, with a small decrease of activation energy. Maximum No.3 is oxygen, the activation energy 204.4 kJ/mol is less than in the first experiments, but it coincides well with the literature data on diffusion. And the maximum number 4 is nitrogen; the reason for the appearance is the same as that of oxygen - poor dynamic vacuum in the chamber.

3.2. Measurements of nitrided samples

For nitriding, the samples were placed in a quartz ampoule, the system was pumped out with a foreline pump, washed three times with nitrogen of technical purity, and after cutting off the pump, the furnace was put into nitriding mode. Three samples from one batch were simultaneously nitrided - the first nitriding of 600 °C - 0.5 h, then the second nitriding of the 2 remaining samples - another 600 °C - 0.5 h. Figure 6 shows the TDIF of nitrided samples:

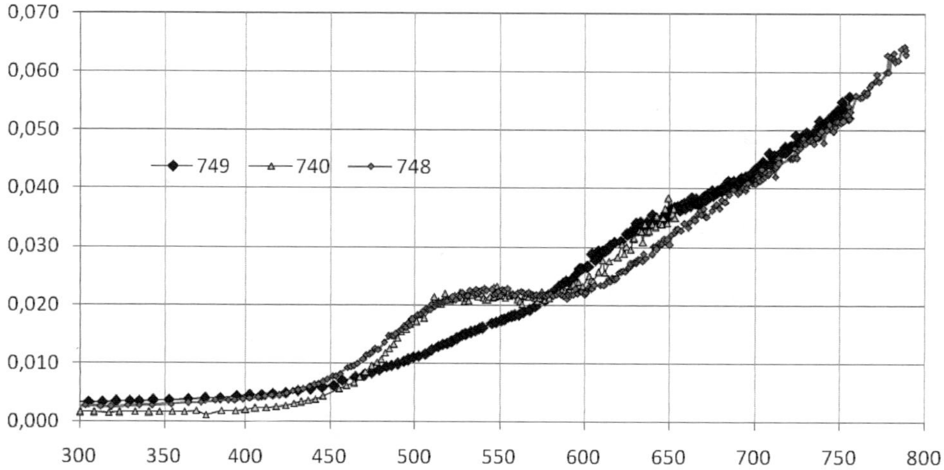

Figure 6. TDIF source and nitrided samples: 740 – initial sample; 748 – sample No. 5, nitrided at 600 °C – 1 h; 749 – remesurement of the same sample No.5.

Further, Figures 7 and 8 show the division into partial maxima of the nitrided samples.

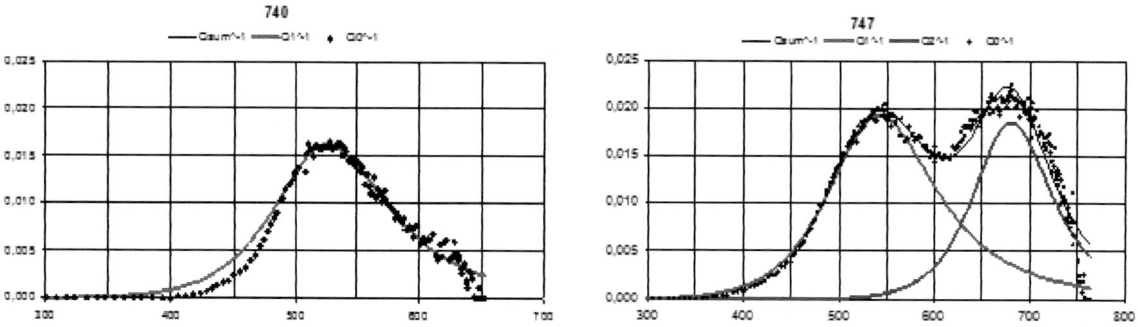

№	T,°C	H, kJ,mol	Q_{max}^{-1}
1	527	196.9	0.01627

№	T,°C	H, kJ/mol	Q_{max}^{-1}
1	542	199.9	0.01929
2	680	235.0	0.01854

Figure 7. Division into partial maxima in experiments 740 and 747: 740 – the original sample of this batch; 747 – sample No. 4, nitrided at 600 °C – 0.5 h.

In these results maximum № 1 is grain-boundary in both experiments; maximum № 2 in 747 is nitrogen.

Figure 8 shows the measurements in the following two experiments.

In experiments 747 and 748, which differ in nitriding time is observed only the grain-boundary (No.1) and nitrogen (No.2) maxima, the parameters of which practically coincide in these experiments. But after re-measuring the sample from experiment 748 (exp 749), an intermediate maximum appeared (No.2 in Figure 8), with an activation energy of 221.8 kJ/mol close to oxygen; its appearance can be explained by oxidation during measurements. The nature of the 4th maximum is unknown.

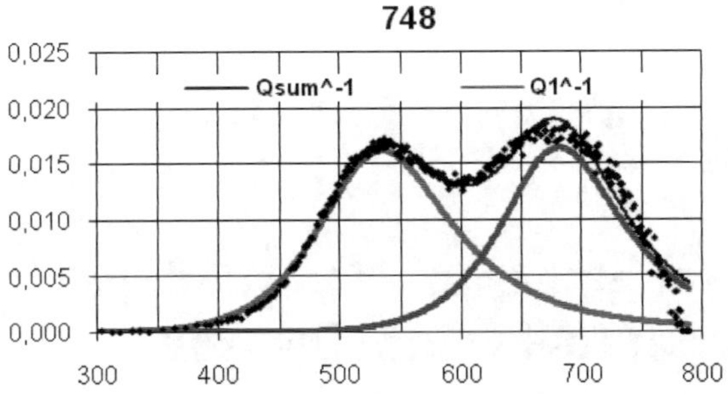

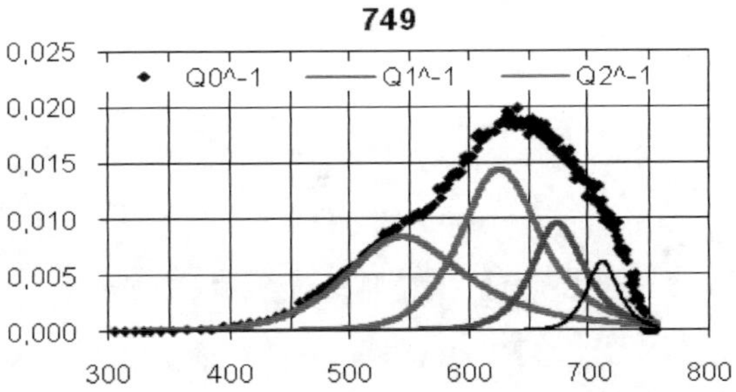

№	T,°C	H, kJ/mol	$Qmax^{-1}$
1	534	198.8	0.01613
2	681	236.2	0.01639

№	T,°C	H, kJ/mol	$Qmax^{-1}$
1	544	201.2	0.00839
2	624	221.8	0.01443
3	673	234.3	0.00956
4	711	244.0	0.00610

Figure 8. Division into partial maxima of experiments 748 and 749: 748 – sample No.5, nitrided at 600 °C – 1 h; 749 – the same sample No.5 when re-measurement.

After additional nitriding, both maxima decreased, but the nitrogen maximum relative to the grain boundary increased, which is logical. A decrease in the height of the grain boundary maximum was also observed in a series of experiments with oxidation in air.

4. Conclusions

It was established that during oxidation in repeated measurements an oxygen maximum appeared, the activation energy decreases in successive experiments: 225, 222, 204 kJ/mol. In repeated measurements of nitrided samples, the oxygen maximum has activation energy 222 kJ/mol. The nitrogen maximum during nitriding has activation energy 234–236 kJ/mol, and the oxygen maximum that appears in these measurements is 225 kJ/mol. After repeated measurements, the carbon maximum appeared with activation energy 165 kJ/mol. The height of the grain boundary maximum in sequential measurements decreases monotonically with an insignificant temperature shift, depending on the measurement conditions, with activation energy from 197 to 203 kJ/mol in different samples.

Acknowledgements

This work was supported by Competitiveness Growth Program of the Federal Autonomous Educational Institution of Higher Education National Research Nuclear University MEPhI (Moscow Engineering Physics Institute).

References

[1] Bisogni E, Mah G and Wert C 1964 Diffusion of gases in special interstitial sites in hafnium. *J. Less-Common Metals* **7** 197–204
[2] Blanter M S, Golovin I S, Neuhäuser H and Sinning H R 2007 *Internal Friction in Metallic Materials. A handbook* (Springer-Verlag) p 540
[3] Mehrer H 1990 *Diffusion in Solids Metals and Alloys* (Springer-Verlag) p 747
[4] Korchagin O N, Kalin B A and Anan'in V M 2011 Hardware-software complex for measuring the characteristics of damped oscillations *Factory laboratory. Diagnostics of materials* **12** (77) 38–42
[5] Wert C and Marx J 1953 A new method for determining the heat of activation for relaxation processes *Acta Metallurgica* **1** 113–115

IOP Conf. Series: Materials Science and Engineering 1005 (2020) 012009 doi:10.1088/1757-899X/1005/1/012009

Features of creating wear-resistant anti-corrosion coatings with a barrier layer on fragments of fuel claddings from E110 o.ch.

B A Kalin[1], A S Yashin[1], P S Dzhumaev[1], D A Safonov[1], E L Korenevsky[1], D A Fedorov[1], V V Novikov[2], V I Kuznetsov[2], P V Fedotov[2], V P Krivobokov[3], S N Yanin[3], A A Mokrushin[4], K K Polunin[4] and V V Uglov[1, 5]

[1] National Research Nuclear University MEPhI (Moscow Engineering Physics Institute), Kashirskoe highway 31, 115409 Moscow, Russia
[2] JSC «High-tech Research Institute of Inorganic Materials named after academician A.A Bochvar», Rogova street, 5a, 123098 Moscow, Russia
[3] National Research Tomsk Polytechnic University, Lenina avenue, 30, 634050 Tomsk, Russia
[4] LUCH Research and Production Association, Research and Development Institute, Federal State Unitary Enterprise, Zheleznodorozhnaya street, 24, 142100 Moscow region, Podolsk, Russia
[5] Belarusian State University, Nezavisimosti avenue, 4, 220030 Minsk, Republic of Belarus

E-mail: yashin_itf@mail.ru

Abstract. Results of the development of protective chromium-containing coatings based on the FeCrNi and CrNi systems for fuel claddings within framework of the accident tolerant fuel (ATF) are presented in this paper. Coatings were deposited by the outer surface of cladding tubes fragments from E110 o.ch. alloy (sponge-based Zr-1%Nb) up to 500 mm length by complex ion-plasma treatment on ILUR-03 and KVK-10 installations. The results of the control tests carried out in high-temperature steam at the GAZPAR bench at 1200 °C up to 400 s showed that Cr-FeCrNi-Cr and Cr-CrNi-Cr coatings reduce total oxygen penetration into the alloy from 144 to 98 and 55 μm, respectively and Cr-CrNi-Cr coatings with a Mo barrier layer completely block the diffusion of oxygen into the material.

1. Introduction

One of the priority tasks of the world energy industry is to increase the lifetime of structural materials in WWER and PWR reactors cores in emergency conditions to increase safety level of its operation. Among the possible ways to improve fuel claddings to create the so-called tolerant fuel (ATF), stable in loss-of-coolant-accidents (LOCA), modification of surface layer of the material and protective coatings deposition by ion-plasma technologies is one of the most effective and technologically method. Researchers from different countries use various elements, alloys and compounds, such as Si, Cr, Al-Zr, FeCrAl, FeCrNi, TiAl, TiN, CrN, ZrN, however, chromium-containing coatings are mentioned as the most promising [1-5].

IOP Conf. Series: Materials Science and Engineering 1005 (2020) 012009 doi:10.1088/1757-899X/1005/1/012009

2. Materials and tools

Fragments of fuel claddings an external diameter of Ø9.1 mm and Ø9.5 mm up to 500 mm length made of alloy E110 o.ch. (sponge-based Zr-1% Nb) were used as samples for this research. Coatings were deposited by complex ion-plasma treatment on the ILUR-03 and KVK-10 installations (NRNU MEPhI, Department No. 9) operating in a semi-automatic mode [6, 7]. The treatment included the sequential execution of a series of operations: preliminary activation of the surface of Ar^+ ion beams E<1 keV; layer-by-layer deposition of coatings using magnetron sputtering systems; intermediate processing of Ar^+ ions with energies up to 35 keV using KVK-10 for surface doping with ion mixing mode [8, 9].

During processing at the ILUR-03 installation, shown at Figure 1 (a), the tubular sample was placed horizontally and sequentially moved through a vacuum chamber with an ion source into a chamber with 3 magnetrons mounted in the radial direction. In the KVK-10 installation, shown at Figure 1 (b), the samples are mounted vertically in a 6-seat cassette, which is placed in a vacuum chamber with 2 ion sources, 3 magnetrons and a high-energy ion implanter mounted on the periphery. The uniformity of processing is ensured by the stable operation of the devices, as well as by the control of the axial rotation speeds and the movement of the samples.

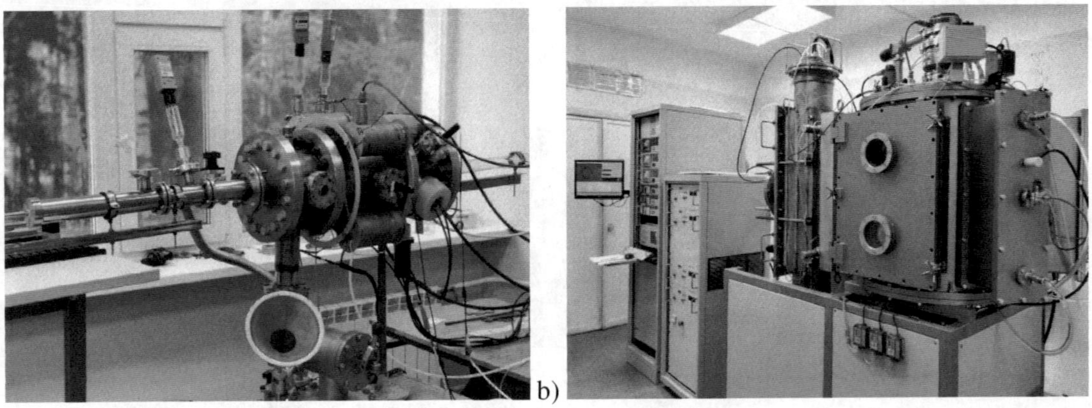

a) b)

Figure 1. Photos of installations for complex ion-plasma surface treatment of materials ILUR-03 (a) and KVK-10 (b).

Corrosion testing of samples were carried out on the GAZPAR stand (LUCH FSUE) by isothermal exposure in water steam at 1200 °C and subsequent cooling of the samples in argon atmosphere at least with 20° $C \cdot s^{-1}$ according to the procedure described in [10]. The corrosion degree was estimated by the specific weight gain and the local oxidation depth (ECR, %), which was calculated as the relative thickness of the zirconium layer, which would turn into oxide, provided that all the absorbed oxygen went into the formation of stoichiometric zirconium oxide ZrO_2.

The surface topography and the transverse structure of the samples in initial state and with coatings before and after high temperature tests were studied in scanning electronic microscopes Carl Zeiss EVO 50 and JEOL JSM-6610LV (resolution up to 3 nm). Both devices are equipped with X-ray microanalysis consoles for investigation of the elemental composition surface of the samples using an energy dispersive detector (EDS, Inca X-act) and a wave dispersive detector (WDS, Inca Wave 500).

3. Results and discussion

As preliminary experiments results shown (Figure 2) chrome coatings are prone to cracking and destruction under the influence of internal stresses arising during high-temperature oxidation in water steam due to thermal expansion and $\alpha \rightarrow \beta$ phase transformation in zirconium at ~800 °C, due to the natural fragility of chromium. Among the disadvantages of chromium coatings, one should also note the relatively high diffusion rate of Cr in Zr and the presence of a eutectic reaction at 1332 °C with, a brittle $ZrCr_2$ phase forming. In this regard, more plastic compositions of FeCrNi and CrNi in various element ratios were used as the basis of coatings in this work.

a)  b) 

Figure 2. The destruction of chromium coatings on samples of alloy E110 o.ch. after high-temperature oxidation in pairs (HTO): a – 800 °C, 5000 s (coating thickness 8-10 µm, SEM image), b – 1000 °C, 5000 s (coating thickness 2-4 µm, optical image).

As can be seen from Figure 3, the coatings deposited had a characteristic multilayer structure. The total thickness of the protective layer in different modes was ~ (2–10) µm.

a) 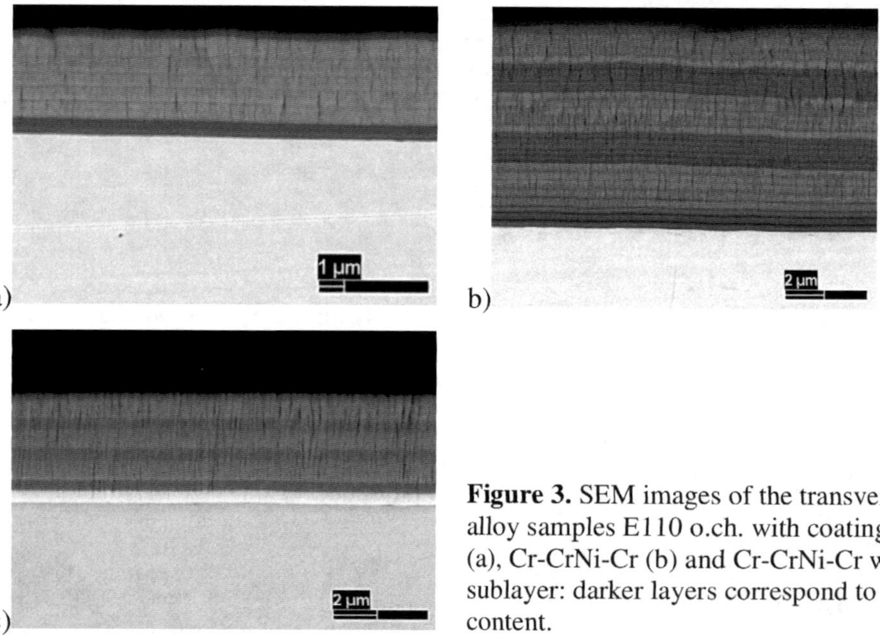

Figure 3. SEM images of the transverse structure of alloy samples E110 o.ch. with coatings Cr-FeCrNi-Cr (a), Cr-CrNi-Cr (b) and Cr-CrNi-Cr with the Mo (c) sublayer: darker layers correspond to an increased Cr content.

Analysis of the transverse structure of samples from alloy E110 o.ch. after oxidation at 1200 °C with 500 s exposure showed that the structural phase state of the material is characterized by the presence of three distinct zones, which is consistent with the known results of other researchers [11]: ZrO_2 oxide zones near the outer and inner surfaces; an intermediate diffusion zone of α-Zr(O), which is essentially a solid solution of oxygen in zirconium stabilized in the α phase; "live" cross-section ex.β-Zr sample not affected by corrosive oxygen.

Oxide film on all samples has homogeneously dark gray color with no visible signs of destruction or shedding. As can be seen from Figure 4, the oxide layer on Zr-1% Nb alloy samples without coatings and with Cr-FeCrNi-Cr coatings has approximately the same thickness of 52±5 µm. There is no thick oxide layer on samples with Cr-CrNi-Cr coatings and Cr-CrNi-Cr coatings with the Mo sublayer which indicates their high protective properties. It was shown that the presence of any coatings under study makes the α-Zr(O) zone structure more uniform and coarse-grained. Its thickness decreases from 85±5 μm for samples without coatings to 44±5 μm for samples with Cr-FeCrNi-Cr coatings and 33±5 μm for samples with Cr-CrNi-Cr coatings. No oxygen diffusion zone α-Zr(O) was detected on the samples with Cr-CrNi-Cr coatings and a Mo sublayer. It should be noted that samples with Cr-FeCrNi-Cr coatings have a weight gain of 12.0±0.5 mg·cm^{-2}, which is larger than that on the samples in the initial state 10.9±0.2 mg·cm^{-2}, which is associated with the active oxidation of coating elements, in particular Fe. Samples with Cr-CrNi-Cr coatings and Cr-CrNi-Cr coatings with the Mo sublayer have a significantly smaller gain of 6.6±0.5 mg·cm^{-2} and 6.1±0.5 mg·cm^{-2}, respectively.

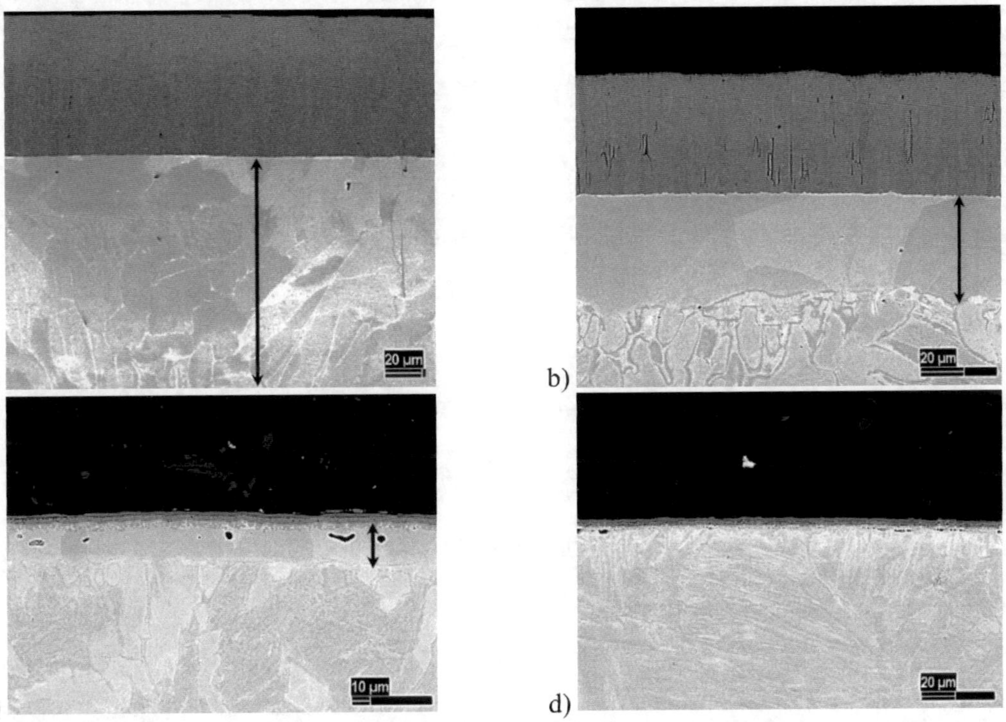

a) b) c) d)

Figure 4. SEM images of the transverse structure of E110 o.ch. samples in the initial state (a) and with the coatings Cr-FeCrNi-Cr (b), Cr-CrNi-Cr (c) and Cr-CrNi-Cr with the Mo sublayer (after corrosion tests 1200 °C 400 s: arrows indicate the α-Zr(O) region.

Elemental analysis of samples with Cr-FeCrNi-Cr coatings showed that the coating atoms are noticeably redistributed over the cross section of the sample as a result of exposure at high temperature in water steam. Fe and Cr atoms shown noticeably dissolving in the alloy under the oxides, along with participation in the formation of $(FeCrNi)O_x$, Cr_2O_3, Fe_mO_n oxides, and thereby slow down the diffusion of oxygen deep into the material, inhibiting oxidation, which can be seen from Figure 5. Moreover, an increased concentration of chromium atoms is observed up to depths of 100-120 μm, and iron - up to 200-300 μm, and then the signal goes to background values.

There are characteristic multiphase structures near the surface of the E110 o.ch. alloy shown at Figure 6, formed as a result of intense diffusion motion of atoms from the Cr-CrNi-Cr and Cr-CrNi-Cr coatings with the Mo sublayer: a dense Cr_2O_3 layer near the outer surface and transition regions with different contents of Zr, Cr, O, Mo (Tables 1 and 2), in particular, $ZrCr_2$ precipitates under that are found.

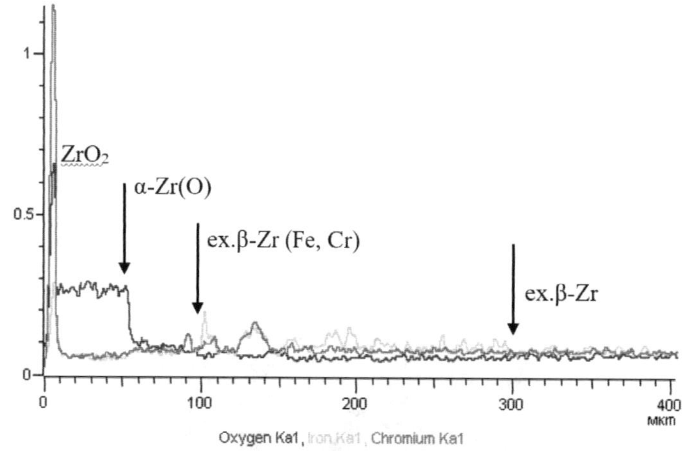

Figure 5. Fe, Cr, O distribution over the depth of a sample coated with Cr-FeCrNi-Cr after oxidation in steam at 1200 °C for 400 s: the arrows indicate the interface between ZrO_2, α-Zr(O), ex.β-Zr and the maximum penetration of Fe atoms (X-ray microanalysis data).

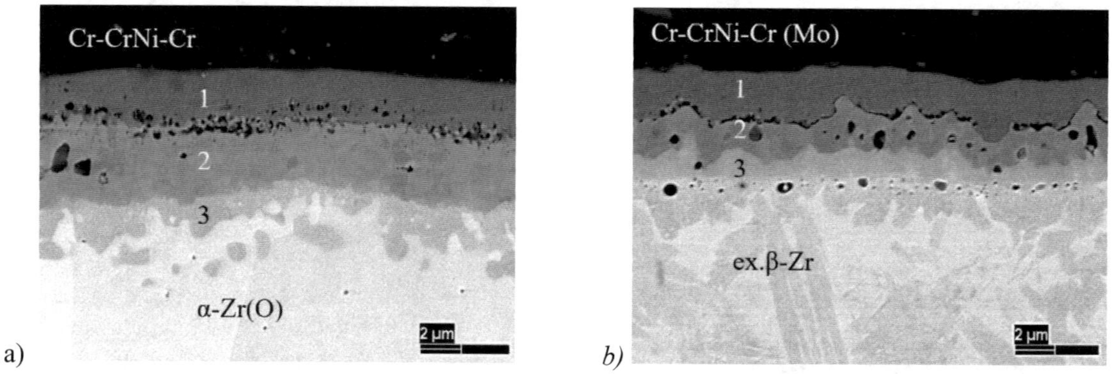

Figure 6. Transverse structure of the E110 o.ch. alloy sample coated with Cr-CrNi-Cr (a) and Cr-CrNi-Cr with the Mo (b) sublayer after the HTO 1200 °C 400 s: the numbers show the elemental composition measurement ranges in tables 1 and 2 respectively.

Table 1. The elemental composition of various regions of E110 o.ch. sample with Cr-CrNi-Cr coating after HTO 1200 °C 400 s.

Measurement area	Element content, %		
	O	Cr	Zr
1 – Cr_2O_3	33.8	65.8	0.0
2 – Cr + Zr	0.0	98.0	2.0
3 – $ZrCr_2$ + O	1.2	50.7	45.9

Table 2. The elemental composition of various regions of E110 o.ch. sample with Cr-CrNi-Cr coating and Mo sublayer after HTO 1200 °C 400 s.

Measurement area	Element content, %			
	O	Cr	Zr	Mo
1 – Cr_2O_3	35.6	64.4	0.0	0.0
2 – Cr + Zr	0.0	96.2	0.6	3.2
3 – Cr + Zr + Mo	0.0	37.9	48.4	13.7

Thus, the Mo barrier layer between E110 alloy and coating Cr-CrNi-Cr inhibits the dissolution of coating elements in sample material and blocks penetration of oxygen from high-temperature steam deep into the cladding effectively, preventing brittle α-Zr(O) phase formation. It became possible to save more than 100 μm of cladding live section in comparison with samples without coatings, which is clearly seen from Figure 7.

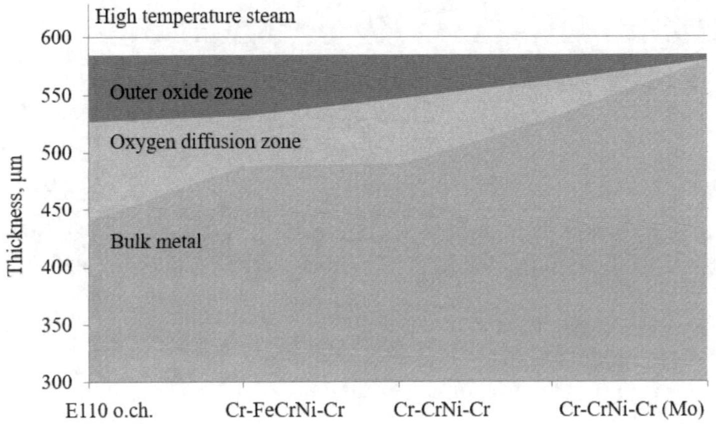

Figure 7. Schematic dependence of the cladding from E110 o.ch. live cross-section, retained after high-temperature oxidation in steam at 1200 °C for 400 s, depending on the type of coating deposited.

4. Conclusion

Protective coatings Cr-FeCrNi-Cr and Cr-CrNi-Cr for cladding tubes from E110 o.ch. alloy, effectively inhibit the oxygen penetration deep into the material during high-temperature tests at 1200 °C with a up to 400 s exposure have been developed. It was shown that use of the Mo barrier layer allows to block oxygen diffusion and prevent the dissolution of coating elements in the sample material, thereby preserving more than 100 μm of the cladding live cross section.

Acknowledgements

This work was supported by Competitiveness Growth Program of the Federal Autonomous Educational Institution of Higher Education National Research Nuclear University MEPhI (Moscow Engineering Physics Institute).

References

[1] Kurt A 2018 *J. of Nuclear Materials* **501** 13–30
[2] Weiwei Xiao, Huiqin Chen, Xiaoshuang Liu et al. 2019 *J. of Nuclear Materials* **526** 151777
[3] Brachet J C, Idarraga I, Le Flem M et al. 2019 *J. of Nuclear Materials* **517** 26–285
[4] Park J.-H, Kim H.-G, Park J.-Y et al. 2015 *Surface and Coatings Technology* **280** 256–259
[5] Bischoff J, Delafoy C, Vauglin C et al. 2018 *Nuclear Engineering and Technology* **50** 223–228
[6] Kalin B A, Volkov N V, Valikov R A et al. 2014 *Bulletin of the Russian Academy of Sciences. Physics* **78** (6) 553–557
[7] Valikov R A, Yashin A S, Yakutkina T V et al. 2015 *J. of Physics: Conf. Ser.* **652** 012068
[8] Yashin A S, Safonov D A, Kalin B A et al. 2017 *IOP Conf. Series: J. of Physics: Conf. Series* **857** 012057
[9] Kalin B A, Volkov N V, Valikov R A et al. 2016 *IOP Conf. Series: J. of Physics: Conf. Series* **130** 012039
[10] Polunin K K, Urusov A A, Mokrushin A A, Soldatkin D M, Kuzma-Kichta Yu A, Kiselev D S, Bespechalov B N and Bazyuk S S 2018 *New in Russian Electrical Power Engineering* **9** 62–71
[11] Polunin K K, Urusov A A, Bazyuk S S, Kiselev D S and Kuzma-Kichta Yu A 2019 *J. Phys.: Conf. Ser.* **1370** 012050

Investigation of a brazed joint EK-181/V/W obtained by Cu-Sn and Cu-Ti amorphous foils

D Bachurina, A Suchkov, Y Gurova and O Sevryukov

National Research Nuclear University MEPhI (Moscow Engineering Physics Institute) Kashirskoe highway 31, Moscow,115409, Russia

E-mail: dmbachurina@mephi.ru

Abstract. The work presents the results of high temperature brazing of reduced activated ferritic martensitic steel EK-181 with pure tungsten, which is essential for DEMO fusion reactor. Vanadium interlayer was used to reduce thermal stresses. Brazing alloys to be used were rapidly quenched into ribbons Cu-12Sn, Cu-20Sn, Cu-12Sn-0,4P for EK-181/V, Cu-50Ti for V/W. Microstructure investigations, mechanical and thermocycling test were carried out.

1. Introduction

The DEMO thermonuclear reactor is the next step towards the creation of an energy thermonuclear reactor; its conceptual projects are currently being developed by all the leading countries of the world. DEMO construction is possible in the 30s of the current centuries, depending on the success of the ITER program. In recent years, there has been intensive development of projects for demonstration and energy fusion reactors: DEMO-C in Russia, ARIES in the USA, SEAFP and PPCS in Europe, SSTR in Japan, etc. [1–4]. Each project is individual in its own way and assumes the presence of the first wall, which is a stressed structural element.

A reduced activation ferritic-martensitic steel (EK-181 [5], Eurofer [6], F82H [7]) is used as a structural material, and tungsten is used as a plasma facing material [8]. Numerous problems limit the direct connection of tungsten and steel due to significant differences in their physical properties, in particular, various coefficients of coefficient of thermal expansion: (10.5-12.3) * $10^{-6}K^{-1}$ in steel and (4.3-6, 0) * $10^{-6}K^{-1}$ at tungsten. This difference causes large residual stresses during cooling, and as a result leads to the destruction of the connection [9,10]. Therefore, firstly, it is necessary to choose the right technology for joining these materials, in this case, the use of brazing technology will ensure the formation of high-quality compounds. Secondly, the use of a buffer layer of pure vanadium will compensate for the differences between steel and tungsten [11,12].

First of all, filler alloys for brazing must meet the requirement of reduced activation. Another important limitation when while joining is the maximum process temperature - it should not exceed 1100°C, because in this case, an active grain growth occurs in a ferritic-martensitic steel.

It follows from the foregoing that the problem of joining dissimilar materials of the DEMO thermonuclear reactor and, in particular, the buffer layer of vanadium alloy and ferritic-martensitic steel is really relevant and has not yet been resolved.

2. Materials and Experimental Methods

In this work the following materials were used:

• Reduced-activated steel of ferritic-martensitic grade EK-181 (Fe – 12Cr – 2W – V – Ta – B, wt.%), In the annealed state, CTE= (10.5-12.3) * $10^{-6}K^{-1}$. Dimensions for mechanical tests: 20x7x7 mm, size for thermocyclic tests: cylinder D = 13mm, h = 6mm;

• tungsten 99.96% purity, CTE= (4.3-6.0) * $10^{-6}K^{-1}$. Dimensions for mechanical tests: 1 x 7 x 7 mm; size for thermocyclic tests D = 10mm, h = 1mm;

• pure vanadium, CTE = (8.3-10.9) * $10^{-6}K^{-1}$. Dimensions for mechanical tests 10x10x0.2 mm; size for thermocyclic tests h = 0.2 μm in the form of a square with a side of ≈10 mm.

Rapidly quenched into ribbon filler alloys STEMET™ were used: Cu-12Sn, Cu-12Sn-0,4P, Cu-20 Sn, Cu -50 Ti, wt.% (the melting temperature: 900 °C, <900 °C, 755 °C, 955-980 °C respectively). STEMET™ brazing alloys are used for production of ITER components [13,14].

In this work, we obtained compounds EK-181 / W (with the presence of spacers from V). In all cases, the V / W joint was obtained using Cu-50Ti brazed, and for EK-181 / V – Cu-Sn. Brazing was carried out in a vacuum furnace at a vacuum not lower than 10^{-5} mm Hg. Comparative analysis of the compounds was carried out on samples that were brazed with the mode №1 1100 ° C, 20 min. A metallographic analysis was carried out, thermocyclic and shear tests were performed for samples:

- EK-181 / Cu-12Sn / V / Cu-50Ti / W,
- EK-181 / Cu-12Sn-0,4P / V / Cu-50 Ti / W,
- EK-181 / Cu-20 Sn / V / Cu-50 Ti / W.

In order to combine the process of brazing and heat treatment of the steel mode №2 was applied 1100 °C, 60 min (to analyze the influence of the holding time during brazing on the microstructure) and mode №3 1100 °C, 60 min (cooling at a rate of 20 °C/min) + 720 °C 180 min , the latter corresponds to the traditional heat treatment of steel EK-181 [15].

The microstructure was investigated on cross sections of brazed cylindrical samples. The microstructure was investigated on an MTI MM500T metallographic microscope, a JEOL JSM-6610LV scanning electron microscope. To identify the microstructure of steel for metallographic studies, a solution was used: 20% HNO_3 + 20% HF + 60% glycerol, etching time 20–40 s.

For thermocyclic tests the samples were placed into a quartz ampoule with a getter from iodide zirconium and evacuated to 10^{-2} mm Hg. Test mode: heating in a tube furnace with resistive heating up to 700 °C followed by quenching in water (30 and 50 cycles). The mode was selected taking into account the operating conditions of the tungsten / steel joint in the "finger" of the HEMJ divertor [16], however, this concept has recently lost relevance, therefore, the task of connecting steel to tungsten is primarily to create the first wall, the operation mode of which is much softer [17]. However, we applied this mode to provide degradation of the joints and proceed fast comparison of different filler alloys as it often used.

We studied the distribution of microhardness $HV_{0.1}$ (load 100 g, holding time 15 s) in the cross section of a brazed seam on a Vickers FM-800 (Future-Tech) microhardness tester. The mechanical shear tests were carried out on an Instron 5569 installation (crosshead speed 1 mm / min) of the samples. Tests were carried out on samples before and after 50 thermal cycles. The test design is presented elsewhere [12].

3. Results and Discussion

The vigorous investigation of V/W microstructure was presented earlier [12]. The microstructure of brazed joints EK-181/V obtained using Cu-Sn filler alloy (Figure 1) consists of the following areas: 1 - zone of ferritic-martensitic steel; 2 - ferrite zone, divided into two areas: a - ferrite formed as a result of carbon depletion, b - ferrite formed as a result of crystallization; 3 - σ (Fe-V); 4 - VC; 5 - pure vanadium. The same microstructures were formed while using Cu-Ge alloys [12]. In the brazed joint EK-181 // Cu-12Sn-0,4P // V ("12Sn + P"), a phosphide is also supposedly formed, near which microcracks form (highlighted by an orange rectangle in Figure 1).

Both in the brazed joint "12Sn" and "12Sn + P", the microhardness changes in a similar way: zone 1 ≈ 400 HV0.1; zone 2 ≈ 200 HV0.1; zones 3 + 4 ≈ 600 HV0.1.

A distinctive feature of the "20Sn" brazed joint is the absence of σ (Fe-V) in the predominant volume of the joint. This phase can only be observed near fillets, which leads to lower microhardness ≈ 400 HV0.1, in comparison with samples in which the sigma phase is detected [12]. When using "12Sn + P" filler alloy, we observe more copper precipitations along grain boundaries.

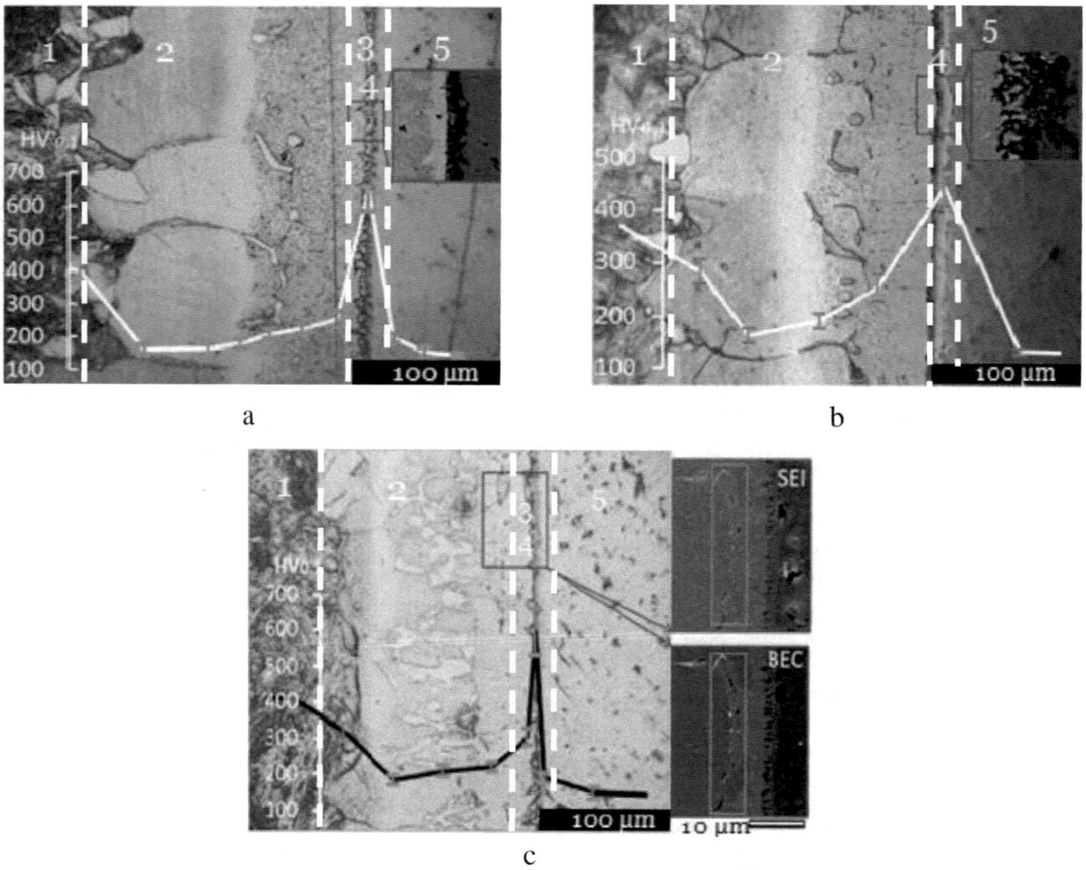

Figure 1. The microstructure of the cross-section of the joint EK-181 / V obtained by mode 1 using filler alloy a) Cu-12Sn; b) Cu-20Sn; c) Cu-12Sn-0.4P.

According to the results of mechanical tests for shear before thermocyclic tests, the tensile strength of the joints is EK-181 / Cu-12Sn / V / Cu-50Ti / W ("12Sn"), EK-181 / Cu-12Sn-0,4P / / V / Cu-50Ti / W ("12Sn + P"), EK-181 / Cu-20Sn / V / Cu-50Ti / W ("20Sn") is 140 MPa, 84 MPa, 160 MPa, respectively (mode 1). The 12Sn and 20Sn compounds are destroyed on the tungsten side (Figure 2 a, b), the 12Sn + P compound (Figure 2 c) is mainly destroyed on the steel side, which is probably associated with the formation of the phosphide phase

There were no changes in the structure of the brazed of these compounds after 30 thermal cycles (Figure 3 a, c; there is no image for "" 12Sn + P). After 50 cycles a structural change in copper precipitates is observed near the fillets which can also indicate a phase change due to the diffusion redistribution of components during heating. In addition, as in the case of compounds obtained using Cu-Ge filler alloy, discontinuities and delamination are formed near the fillets. There is a change in structure from the side of tungsten (Figure 3 b, g). Mechanical tests after 50 cycles give the following result: "12Sn" - 35 MPa, "12Sn + P" - 28 MPa, "20Sn" - 46 MPa.

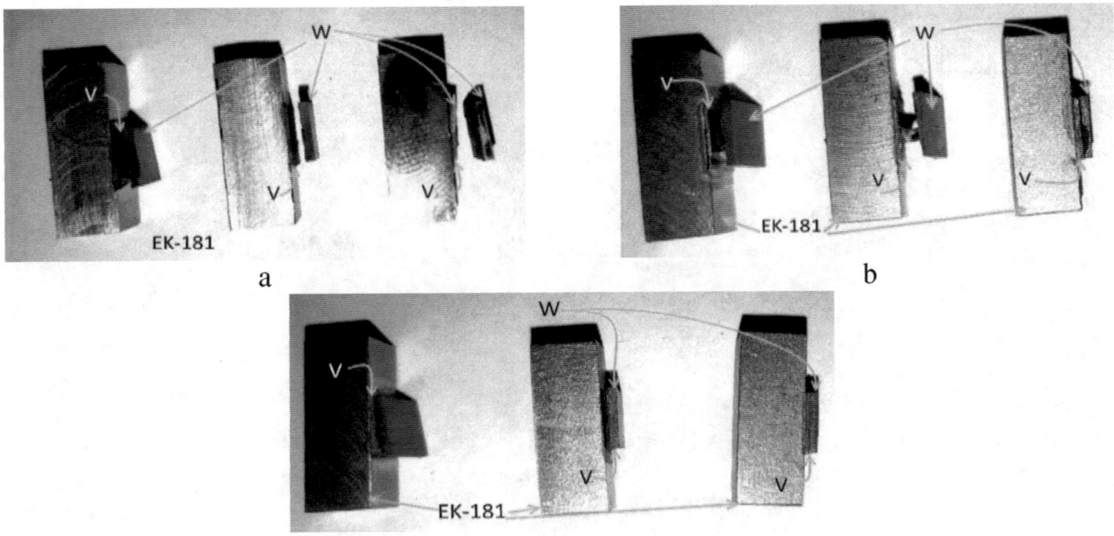

Figure 2. Samples after mechanical tests before thermal cycling:
a) EK-181 / Cu-12Sn / V / Cu-50Ti / W; b) EK-181 / Cu-20Sn / V / Cu-50Ti / W;
c) EK-181 / Cu-12Sn-0,4P / V / Cu-50Ti / W.

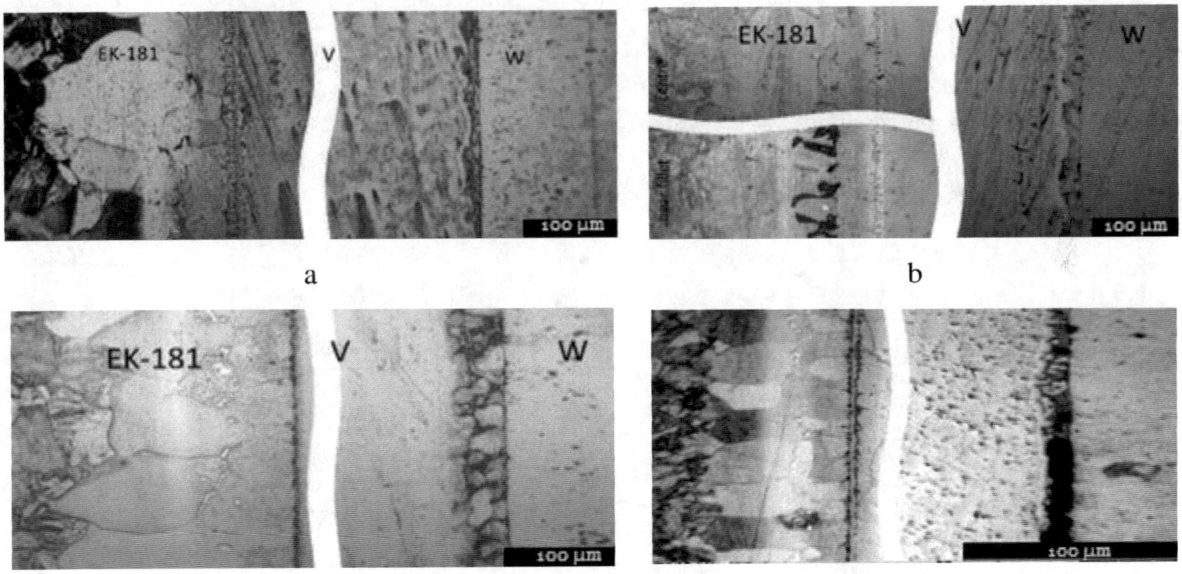

Figure 3. The microstructure of the cross section of the seam of brazed joints:
EK-181/Cu-12Sn /V/ Cu-50Ti /W a– after 30 thermal cycles; b– after 50 thermal cycles;
EK-181 / Cu-20Sn / V / Cu-50Ti / W c– after 30 thermal cycles; g - after 50 thermal cycles.

Based on the described results due to the higher value of shear strength as well as a reduced amount of sigma phase, Cu-20Sn filler alloy was considered for further studies.

To combine the process of brazing and heat treatment of steel an analysis was made of the dependence of the microstructure on the temperature-time regime as well as a change in mechanical properties during the transition to the heat treatment of steel. Comparing the microstructure of the brazed EK-181 / V (figure 4) of the samples obtained in modes 2, 3 with the sample (mode 1), zones with and without sigma phase are observed. No grain growth of ferritic zine nor growth of other phases were observed.

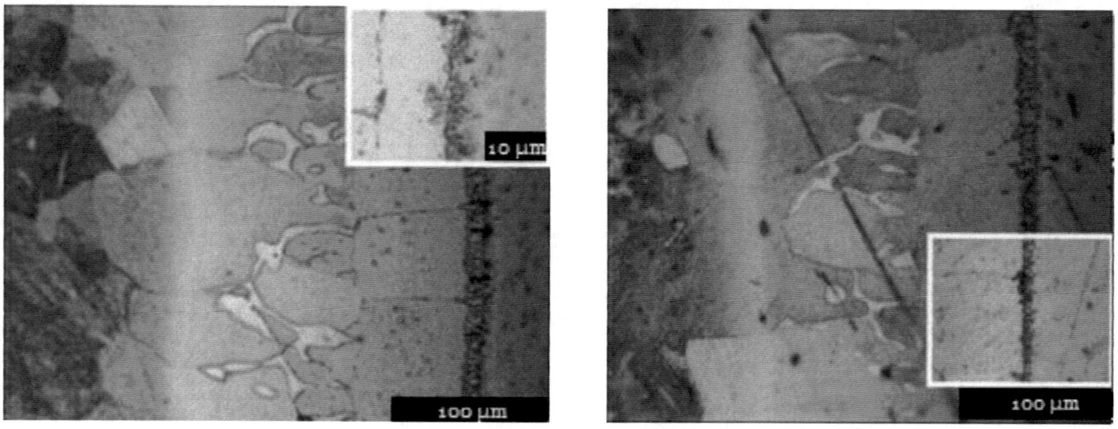

a b

Figure 4. Brazed seam EK-181 / V obtained using Cu-20Sn according to the modes: a) mode 2; b) mode 3.

Final tensile strength of the brazed joint obtained by mode 3 - 93 ± 24 MPa. The destruction of the samples occurs on the side of tungsten.

Figure 5 presents the summary results of the mechanical tests carried out in the work. It is noted that thermocyclic tests lead to a decrease in strength by more than a factor of two, which is associated with delamination and chipping of the phases on the V / W side. The heat treatment of steel leads to a slight decrease in strength, However, the achieved values are higher than in some works [18,19], as well is while using Cu-Ge alloys [10,12].

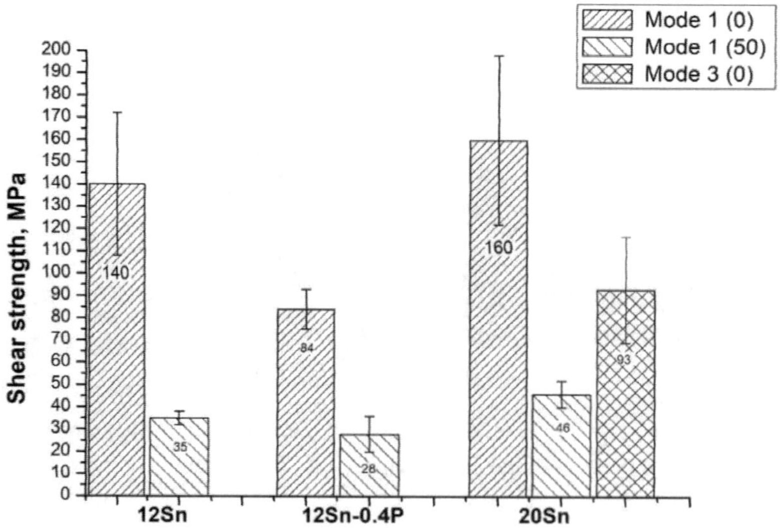

Figure 5. Results of mechanical shear tests

4. Conclusion

High-temperature brazing of the EK-181 / V / W joint was carried out with rapidly quenched into ribbon filler alloys: Cu-50Ti for V / W and Cu-12Sn, Cu-20Sn, Cu-12Sn-0.4P for EK-181 / V in three modes:

№ 1. 1100 °C, 20 min (for comparative analysis);

№ 2. 1100 °C, 60 min (to analyze the effect of the holding time during brazing on the microstructure);

№ 3. 1100 °C, 60 min + 720 °C 180 min (corresponds to the heat treatment mode EK-181).

The brazed seam EK-181 / V consists of a zone of ferritic-martensitic steel, a zone of ferrite, divided into two areas: a - ferrite formed as a result of carbon depletion, b - ferrite formed as a result of crystallization, vanadium carbide, σ(Fe-V), pure vanadium. However, the amount of σ (Fe-V) in the brazed joint obtained by Cu-20Sn is much lower compared to other brazing alloys. The absence of this phase results in a higher shear strength of the joint.

It was established that an increase in the holding time during brazing doesn't lead to the microstructural changes. Subsequent annealing at 720 ° C for 180 min also does not affect.

Mechanical shear tests were carried out for brazed joints obtained according to mode 1 (before thermocyclic and after 50 thermal cycles of 700 ° C - quenching in water). Also shear strength was investigated for as-joined "20Sn" obtained by mode 3 (before thermocyclic tests).

The strength of the joint EK-181 / Cu-20Sn / V / Cu-50Ti / W is:
Mode 1 (0 thermal cycles) – 160 ± 38 MPa;
Mode 1 (50 thermal cycles) – 46 ± 6 MPa;
Mode 3 (0 thermal cycles) – 93 ± 24 MPa.
The strength of the joint EK-181 / Cu-12Sn / V / Cu-50Ti / W is:
Mode 1 (0 thermal cycles) – 140 ± 32 MPa;
Mode 1 (50 thermal cycles) – 35 ± 3 MPa;
The strength of the joint EK-181 / Cu-12Sn-0,4P / V / Cu-50Ti / W is:
Mode 1 (0 thermal cycles) – 84 ± 9 MPa;
Mode 1 (50 thermal cycles) – 28 ± 8 MPa.

Acknowledgements
The work was acknowledged by The Council for grants of President of Russian Federation.

References
[1] T R Barrett et al 2016 *Fusion Eng. Des.* **109111** 917
[2] M I Subbotin, K D K, and F E A 2010 *Quest. Nucl. Sci. Technol.* **3** 55 (In Russian)
[3] K Tobita et al 2018 *Fusion Eng. Des.* **136** 1024
[4] G Federici et al 2018 *Fusion Eng. Des.* **136** 729
[5] V M Chernov, M V. Leont'eva-Smirnova, E M Mozhanov, N S Nikolaeva, A N Tyumentsev, N A Polekhina, I Y Litovchenko and E G Astafurova 2016 *Tech. Phys.* **61** 209
[6] A A F Tavassoli et al 2004 *Materials design data for reduced activation martensitic steel type EUROFER* (North-Holland) p 257
[7] T Hirose, T Kato, H Sakasegawa, H Tanigawa and T Nozawa 2020 *Fusion Eng. Des.* **160** 111823
[8] V Philipps 2011 *J. Nucl. Mater.* **415** S2
[9] W W Basuki and J Aktaa 2011 *J. Nucl. Mater.* **417** 524
[10] D Bachurina, A Suchkov, B Kalin, O Sevriukov, I Fedotov, P Dzhumaev, A Ivannikov, M Leont'eva-Smirnova and E Mozhanov 2018 *Nucl. Mater. Energy* **15** 135
[11] W W Basuki and J Aktaa 2012 *J. Nucl. Mater.* **429** 335
[12] D Bachurina, A Suchkov, A Filimonov, I Fedotov, M Savelyev, O Sevryukov and B Kalin 2019 *Fusion Eng. Des.* **146** 1343

IOP Conf. Series: Materials Science and Engineering 1005 (2020) 012010 doi:10.1088/1757-899X/1005/1/012010

[13] B A Kalin, A N Suchkov, V T Fedotov, O N Sevryukov, A A Ivanniko and A A Gervash 2016 *Nucl. Mater. Energy* **9** 388
[14] I V. Mazul, V A Belyakov, A A Gervash, R N Giniyatulin, T Guryeva, V E Kuznetsov, A N Makhankov, A A Okunev and O N Sevryukov 2016 *Fusion Eng. Des.* **109–111** 1028
[15] A G Ioltukhovskii, M V Leont'eva-Smirnova, V M Chernov, V V Tsvelev, M I Solonin, V N Golovanov and V K Shamardin 2002 *Termicheskaya Obrab. Met.* **44** 11
[16] P Norajitra, S Antusch, R Giniyatulin, V Kuznetsov, I Mazul, H J Ritzhaupt-Kleissl and L Spatafora 2011 *Fusion Eng. Des.* **86** 1656
[17] Y Huang, F Cismondi, E Diegele, G Federici, A Del, F Moro and N Ghoniem 2018 *Fusion Eng. Des.* **135** 31
[18] J C Wang, J Huang, H Sun, X Gao, W Wang, Q Li, C Xie, X Wang, Z Chen, Q Gao, S Liu, G N Luo and J Li 2019 *Fusion Eng. Des.* **138** 313
[19] J Wang, W Wang, R Wei, X Wang, Z Sun, C Xie, Q Li and G Luo 2017 *J. Nucl. Mater.* **485** 8

Study of Ti-Zr-Nb-Be filler metal interaction with silicon carbide based ceramics during brazing process

A A Eroshenko, A A Ivannikov, I V Fedotov, A N Suchkov, P S Dzhumaev and O N Sevryukov

National Research Nuclear University MEPhI (Moscow Engineering Physics Institute) Kashirskoe highway 31, Moscow,115409, Russia

Email: nastyaerosh1@yandex.ru

Abstract. This work is devoted to an analysis of the composition of a silicon carbide based ceramic brazed seam as a result of its interaction with rapidly quenched titanium-zirconium-niobium-beryllium filler metal. Structural-phase studies based on EDX and EBSD analysis, mechanical shear tests, and microhardness measurements of brazed joints were carried out. It was shown that titanium, zirconium, and niobium silicides as well as particles of titanium carbosilicides and silicon carbides in the silicon matrix are formed in the brazed seam, probably because of the presence of free silicon in the base material, which leads to increasing joint microhardness and unstable shear strength results.

1. Introduction

Ultrafast solidification technology makes it possible to obtain flexible ribbons and powders from hard-deformed ingots that have a range of advantages over their crystalline counterparts: they have a uniform phase state throughout the volume, narrow melting and solidification intervals, and high adhesive and capillary activity. Rapidly quenched filler metals have proven themselves well for dissimilar material brazing [1] and for creating ceramic–ceramic and ceramic–metal joints that can withstand mechanical stresses without noticeable deterioration of their properties [2–4].

Silicon carbide has great technological importance in modern industries and applications in different areas due to its high chemical resistance and thermal stability. It shows promise for use as the basis of a composite material in generation IV thermal neutron reactors. The shells of fuel elements made of ceramic based on silicon carbide have a lower value of thermal neutron capture cross-section than classical zirconium alloys and also show increased radiation resistance when irradiated [5]. The application of silicon carbide based ceramic in pressurized-water reactors (PWRs) as fuel element cladding will provide a radical increase in the radiation safety of modern nuclear power plants in line with the concept of accident-tolerant nuclear fuel that is resistant to accidents with loss of coolant (LOCA) [6]. However, after some manufacturing methods, silicon carbide may contain additives that will greatly effect the fuel element shell operation and the brazing process itself. One such additive is free silicon, which often remains after silicon carbide reaction bonding processes [7,8]. In this regard, study of the brazing process in the presence of free silicon in silicon carbide is an urgent task.

Brazing of silicon carbide can be carried out with filler metals based on gold or silver with titanium additives [9,10]. Unfortunately, due to the high cost of gold and silver, it is promising to use filler metals based on titanium, for example, in Ref. [11], where silicon carbide was brazed using active solder Ti-24 at.% Ni with zirconium additives to reduce residual thermal stresses. Good results for the mechanical shear strength of brazed joints of Cf/SiC composite were obtained in Ref. [12], where filler metal of the Ti-Zr-Be system was used. It was shown that a brazed seam consists of possible phases of carbides, silicides, and carbosilicides of titanium and zirconium, which has a high melting point. The presence of high-temperature melting point phases in the brazed seam has great importance because the LOCA accident temperature can rise up to 1200 °C [13]. The authors of Ref. [14] note that the most promising results for high-temperature applications of silicon carbide joints can be achieved with filler metals based on titanium.

The aim of this work is to study the interaction of Ti-40Zr-8.5Nb-1.5Be filler metal with silicon carbide ceramics containing free silicon during brazing, to analyse the structural-phase state of brazed joints, and to evaluate the mechanical characteristics.

2. Materials and Experimental Methods

Ingots of Ti-40Zr-8.5Nb-1.5Be (w.p.%) filler metal were produced by the argon-arc melting method in an MIFI-9 furnace from pure components: titanium and zirconium, NB1 alloy, and Ti-6Be ligature. The obtained ingots were used for filler metal ribbon production on Crystal-702 installation in a helium atmosphere. The fabricated filler metal tape had a thickness of 50 ± 10 μm and full melting temperature of 1280 °C.

As the main material in the work, samples from ceramics based on silicon carbide, in which the volume fraction of free silicon was 18.5% and the free porosity rate was 2%, were used. Silicon carbide ceramic samples for microstructural studies and mechanical tests were brazed in an SShVE-1.25/25 vacuum furnace with the use of Ti-40Zr-8.5Nb-1.5Be filler metal at a temperature of 1420 °C for holding times of 1, 5, and 20 min. in a vacuum at $\sim 8 \cdot 10^{-5}$ torr. Such a high brazing temperature was chosen in order to obtain the most stable and refractory phase composition in the brazed joints. After brazing, the joints were cooled at a rate of 5 °C/min. to a temperature of 600 °C and then cooled with a speed of furnace cooling.

Microstructure images of the silicon carbide ceramic brazed joints were obtained in the backscattered electron mode using a Carl Zeiss EVO 50 scanning electron microscope. Elemental analysis was performed by INCA X-Act EDX analyzer. The crystal lattice identification of the phases formed in the brazed seam was performed by an EBSD NORDLYS detector. The relative fraction of porosity in the brazed joint was estimated by ImageJ image processing software. The image obtained at a magnification of 100× was graphically processed and the relative fraction of pores was calculated by relation of the pore area to the area of the brazed seam. The Gibbs energy of the reactions during brazing was calculated in the HSC6 software package per 1 mol of interacting silicon/carbon/silicon carbide at a brazing temperature of 1420 °C.

The shear strength test of silicon carbide brazed joints obtained with Ti-40Zr-8.5Nb-1.5Be filler metal was carried out on a Quazar 50 testing machine with a traverse speed of 1 mm/min. The Vickers microhardness of the joint was measured on an HVS1000 machine with a load of 300 grams and a holding time of 30 sec.

3. Research results and discussion

3.1. Study of the structural-phase state of silicon carbide ceramic brazed joint obtained with Ti–40Zr–8.5Nb–1.5Be filler metal

The brazed joints of silicon carbide ceramics obtained with Ti-40Zr-8.5Nb-1.5Be filler metal are shown in Figure 1.

It is noted that in the brazed joint and main material of silicon carbide ceramic, pores were formed. The size and quantity of pores depend on the exposure time at the brazing temperature. The relative fraction of pores in the brazed joint is shown in Table 1. The brazed joint thickness was 50 μm for holding times of 1 and 5 min. and 80 μm for a holding time of 20 min.

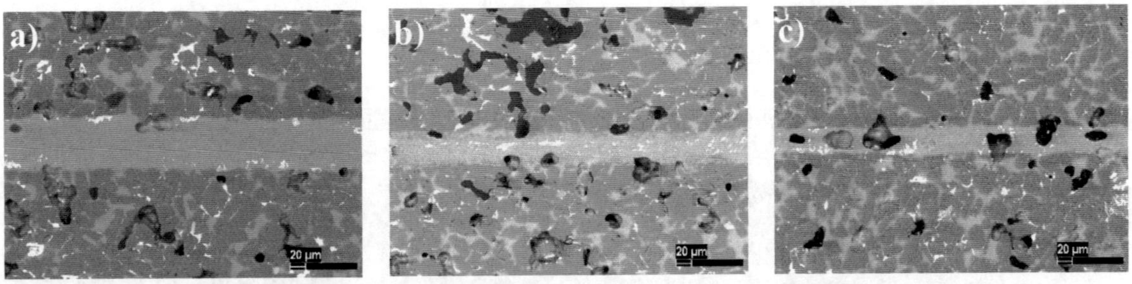

Figure 1. Microstructure of silicon carbide ceramic brazed joints obtained with rapidly quenched filler metal Ti-40Zr-8.5Nb-1.5Be at a temperature of 1420 °C and brazing times of (a) 20 min, (b) 5 min, and (c) 1 min.

Table 1. Relative fraction of pores in silicon carbide ceramic brazed joints.

Brazing mode	Percentage porosity
1420 °C – 1 min.	18.0
1420 °C – 5 min.	10.6
1420 °C – 20 min.	9.7

In Figure 2, a fragment of brazed seam is shown with the results of EDX and EBSD analysis, while Figure 3 shows an X-ray map of the whole joint.

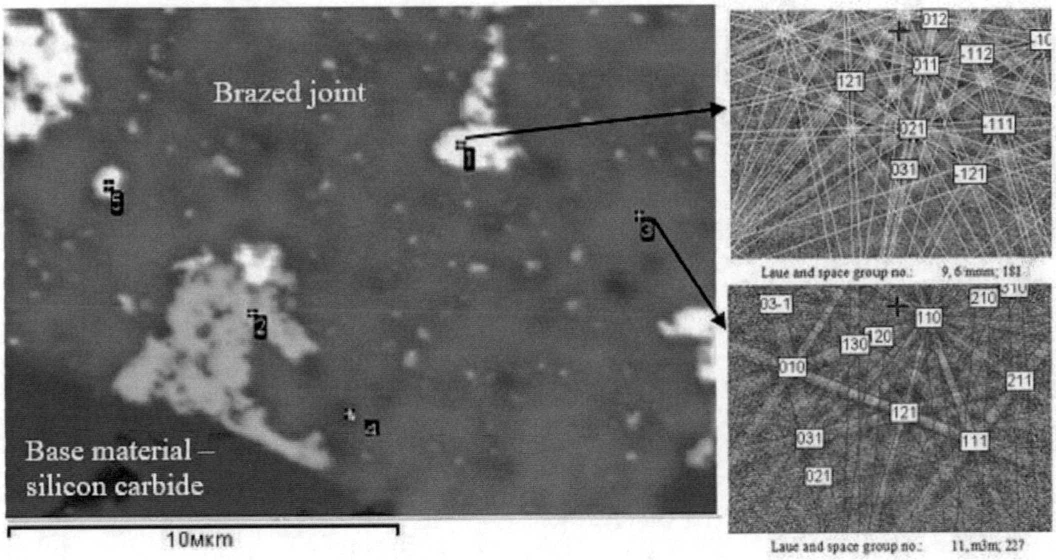

Zone	C	Si	Ti	Zr	Nb	Space group, crystal lattice	Possible phases
1	0	67.7	18.3	0	14.0	181, hexagonal	$(Ti,Nb)Si_2$
2	14.6	58.8	25.2	0.8	0.6	–	(Ti_3SiC_2, Ti_2SiC)
3	0	100	0	0	0	227, cubic (diamond type)	Si
4	47.4	52.6	0	0	0	–	–
5	0	72.3	19.1	8.6	0	–	$(Ti,Zr)Si_2$

Figure 2. EDX (a.p.%) and EBSD microanalysis of silicon carbide ceramic brazed joints obtained with Ti-40Zr-8.5Nb-1.5Be filler metal and a brazing mode of 1420 °C for 20 min.

Figures 2 and 3 show that the brazed seam consists of a silicon matrix (diamond-type lattice) and inclusions of mainly silicides of titanium, zirconium, and niobium as well as silicon carbide particles smaller than 1 μm in size. The presence of some amount of carbon-rich phase-carbosilicides (possibly Ti_3SiC_2 or Ti_2SiC) in the interface of the brazed seam and the base material was also found.

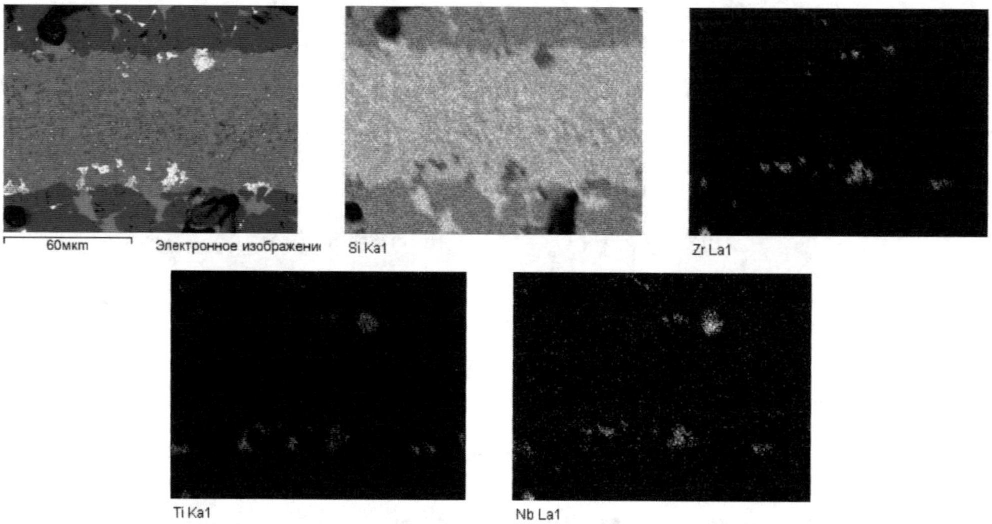

Figure 3. EDX mapping of silicon carbide ceramic brazed joint obtained with Ti-40Zr-8.5Nb-1.5Be filler metal and a brazing mode of 1420 °C for 20 min.

According to the EBSD analysis, titanium-zirconium-niobium silicides crystallize in the hexagonal lattice corresponding to the $NbSi_2$ compound and have sizes from 1 to 30 μm. Titanium, zirconium, and niobium silicides ($ZrSi_2$, $TiSi_2$, $NbSi_2$) as well as titanium carbosilicides and pure silicon have high melting points (more than 1200 °C), which are suitable for high temperature applications. The formation of silicides with a silicon content of 66% and chemical formula of $(Ti,Zr,Nb)Si_2$ is probably caused by a large excess of silicon in the base material, since it was calculated that the formation of a silicide with a higher content of titanium or zirconium is more thermodynamically beneficial (1)–(4):

$$Ti + 2Si \rightarrow TiSi_2 \qquad \Delta G = -79.3\ kJ/mol\ (T = 1420\ °C) \qquad (1)$$

$$5Ti + 3Si \rightarrow Ti_5Si_3 \qquad \Delta G = -195.3\ kJ/mol\ (T = 1420\ °C) \qquad (2)$$

$$Zr + 2Si \rightarrow ZrSi_2 \qquad \Delta G = -70.3\ kJ/mol\ (T = 1420\ °C) \qquad (3)$$

$$5Zr + 3Si \rightarrow Zr_5Si_3 \qquad \Delta G = -187.5\ kJ/mol\ (T = 1420\ °C) \qquad (4)$$

The probable mechanism of $(Ti,Zr,Nb)Si_2$ formation consists in the following silicide formation reaction with formulas (5), (6) starting with reactions (2), (4):

$$Ti_5Si_3 + 2Si \rightarrow 5TiSi \qquad \Delta G = -26.1\ kJ/mol\ (T = 1420\ °C) \qquad (5)$$

$$TiSi + Si \rightarrow 5ZrSi_2 \qquad \Delta G = -31.0\ kJ/mol\ (T = 1420\ °C) \qquad (6)$$

Also the presence of an excess of free silicon in the brazed ceramic affects the phase composition of the brazed joint, as it was expected that during the interaction of filler metal with the silicon carbide, the formation of carbides will occur according to the possible formulas (7-8):

$$Ti + SiC \rightarrow TiC + Si \qquad \Delta G = -105.6\ kJ/mol\ (T = 1420\ °C) \qquad (7)$$

$$Zr + SiC \rightarrow ZrC + Si \qquad \Delta G = -121.7\ kJ/mol\ (T = 1420\ °C) \qquad (8)$$

However, as shown in Ref. [12], titanium and zirconium carbides can be formed during silicon carbide brazing, but the base material should consist not of free silicon but of carbon, which makes possible the implementation of the reaction according to formulas (9) and (10).

$$Ti + C \rightarrow TiC \qquad \Delta G = -163.4\ kJ/mol\ (T = 1420\ °C) \qquad (9)$$

$$Zr + C \rightarrow ZrC \qquad \Delta G = -179.5\ kJ/mol\ (T = 1420\ °C) \qquad (10)$$

Thus the obtained phase composition of the brazed seam can be primarily described by the interaction of filler metal with free silicon and the formation of mainly silicides, while the formation of carbides is less thermodynamically beneficial and is blocked by the lower Gibbs energy (11).

$$Ti + 2SiC \rightarrow TiSi_2 + 2C \qquad \Delta G = -21.5\ kJ/mol\ (T = 1420\ °C) \qquad (11)$$

The small silicon carbide particles in the brazed seam are related to the good solubility of carbon in liquid silicon (the brazing temperature is higher than the melting point of silicon) and the complete absence of solubility in solid, which leads to the formation of silicon carbide particles during crystallization, while the porosity is probably associated with large areas of silicides-brittle inclusions in the brazed joint and their subsequent chipping from the silicon matrix of the brazed joint and the base material during sample preparation for the research.

3.2. Study of the silicon carbide ceramic joint mechanical characteristics obtained with Ti–40Zr–8.5Nb–1.5Be filler metal

The microhardness and shear strength results of silicon carbide ceramic brazed joint obtained with rapidly quenched Ti-40Zr-8.5Nb-1.5Be filler metal are shown in Table 2.

Table 2. Average microhardness and shear strength of silicon carbide ceramic brazed joints.

Brazing mode	Average microhardness of the brazed joint, GPa	Average value of brazed joint shear strength, MPa
1420 °C – 1 min.	12.2 ± 0.4	22.5 ± 9.3
1420 °C – 5 min.	11.3 ± 0.4	37.8 ± 19.2
1420 °C – 20 min.	12.5 ± 0.6	24.5 ± 9.9

The average microhardness of brazed joints ranged from 11.3 GPa (with a brazing holding time of 5 min.) to 12.5 GPa (with a holding time of 20 min.) due to the presence of hard particles of titanium, niobium silicides, titanium carbosilicides, and silicon carbides in the silicon matrix of the brazed joint.

According to the shear strength results (Table 2), the silicon carbide ceramic brazed joints have a shear strength of 22.5 (with a holding time of 1 min.) to 37.8 MPa (with a holding time of 5 min.). Large scatter in the shear strength results is connected with the high average microhardness as well as the presence of fragile silicon and silicide phases in the brazed seam.

Conclusions

An analysis of the interaction of Ti-40Zr-8.5Nb-1.5Be filler metal with silicon carbide ceramic consisting of free silicon during the brazing process was carried out. The study of the structural-phase state of brazed joints obtained at a temperature of 1420 ° C and holding times of 1, 5, and 20 min. is presented. It is shown that the brazed joint consists of pure silicon with inclusions of titanium, zirconium, and niobium silicides as well as titanium carbosilicides and silicon carbides, due to the presence of free silicon in the base material. No formation of fusible inclusions was found in the compounds, but the resulting brazed joint has high microhardness due to the presence of fragile silicide inclusions in the silicon matrix, which leads to a large scatter in the shear strength results. Increasing the exposure time to 20 min. leads to a decrease in the porosity level with an increase in the thickness of the brazed joint from 50 to 80 μm.

Acknowledgements

This work was supported by Competitiveness Growth Program of the Federal Autonomous Educational Institution of Higher Education National Research Nuclear University MEPhI (Moscow Engineering Physics Institute).

References

[1] Gorynin I V, Farmakovsky B V, Vasiliev A F, Vinogradova T S and Samodelkin E A 2017 Activated soldering of dissimilar materials with amorphous solders *Inorg. Mater. Appl. Res.* **8** 892–8

[2] Fedotov I V, Suchkov A N, Fedotov V T, Sevryukov O N, Kalin B A and Ivannikov A A 2015 Brazing of hexagonal boron-nitride ceramics with VT1-0 titanium alloy using a rapidly quenched titanium-based brazing alloy *Weld. Int.* **29** 222–6

[3] Fedotov I V, Suchkov A N, Sevryukov O V, Ivannikov A A, Gasnikov P S and Sannikova S N 2017 Brazing of zirconia-based ceramics to 29NK alloy using CTEMET 1202 amorphous tape brazing alloy *Weld. Int.* **31** 979–83

[4] Zou J, Jiang Z, Zhao Q and Chen Z 2009 Brazing of Si3N4 with amorphous Ti40Zr25Ni15Cu20 filler *Mater. Sci. Eng. A* **507** 155–60

[5] Stempien J D, Carpenter D M, Kohse G and Kazimi M S 2013 Characteristics of composite silicon carbide fuel cladding after irradiation under simulated PWR conditions *Nucl. Technol.* **183** 13–29

[6] Deck C P, Jacobsen G M, Sheeder J, Gutierrez O, Zhang J, Stone J, Khalifa H E and Back C A 2015 Characterization of SiC-SiC composites for accident tolerant fuel cladding *J. Nucl. Mater.* **466** 667–81

[7] Evans R S, Bourell D L, Beaman J J and Campbell M I 2003 Reaction bonded silicon carbide: SFF, process refinement and applications *Proc. Solid Free. Fabr. Symp.* 414–22

[8] Zhang N L, Yang J F, Deng Y C, Wang B and Yin P 2019 Preparation and properties of reaction bonded silicon carbide (RB-SiC) ceramics with high SiC percentage by two-step sintering using compound carbon sources *Ceram. Int.* **45** 15715–9

[9] He H, Lu C, He Y, Zheng W, Yang J, Wang L, Sun Y and Gao Z 2019 Characterization of SiC ceramic joints brazed using Au-Ni-Pd-Ti high-temperature filler alloy *Materials (Basel).* **16**

[10] Liu Y, Huang Z R and Liu X J 2009 Joining of sintered silicon carbide using ternary Ag-Cu-Ti active brazing alloy *Ceram. Int.* **35** 3479–84

[11] Qi Q, Zhanga J, and Hu H 2019 Benefits of Zr additive element in the Ti–24Ni eutectic filler in vacuum brazing of SiC ceramics *Vac. 162 110–113*

[12] Fan D, Huang J, Sun X, Yang J, Zhao X and Chen S 2016 Correlation between microstructure and mechanical properties of active brazed Cf/SiC composite joints using Ti-Zr-Be *Mater. Sci. Eng. A*

[13] Ali R, Mansur T M N T, Baharudin N H and Hassan S I S 2016 *Environmental impacts of renewable energy* (Elsevier Inc.)

[14] Asthana R, Singh M, Lin H T, Matsunaga T and Ishikawa T 2013 Joining of SiC fiber-bonded ceramics using silver, copper, nickel, palladium, and silicon-based alloy interlayers *Int. J. Appl. Ceram. Technol.* **10** 801–13

Manufacturing of high entropy alloy in the Ni-Nb-Co-Fe-Cr system by rapidly solidification method for oxide ceramic brazing

A I Gabov, A A Ivannikov and O N Sevryukov

National Research Nuclear University MEPhI (Moscow Engineering Physics Institute) Kashirskoe highway 31, Moscow,115409, Russia

E-mail: artem.g1297@yandex.ru

Abstract. This paper presents results of research of high-entropy alloys of the Ni-Nb-Co-Fe-Cr system in the as-cast state and after rapid quenching from the melt. The results of experiments on obtaining a brazed joint of aluminum-oxide ceramics with using the filler metal of this system and the results of a studying this joint are presented. Differential thermal analysis (DTA), X-ray diffraction analysis (XRD), scanning electron microscopy (SEM), energy-dispersive X-ray microanalysis (EDX) and measurement of microhardness were used as experimental research methods. The liquidus and solidus temperatures of high-entropy alloy $Nb_{0.73}CoCrFeNi_{2.1}$ were determined. The phase composition of alloys of the Ni-Nb-Co-Fe-Cr system was studied. It was shown, that alloys of this system can be used as filler metal to create a joint between ceramics. The microhardness of the brazed seam was studied. Based on these results the brazing mode for samples of aluminum-oxide ceramics with using high-entropy alloy $Nb_{0.73}CoCrFeNi_{2.1}$ as a filler metal was chosen.

1. Introduction

In modern technologies, there is often a need to create joints with between two or more structural elements. One way to create one-piece joints is brazing. Brazing is the process of connecting materials in a solid state with filler metal that wets the joined surfaces at the brazing temperature, fills the gap between them and forms a brazed joint during crystallization [1]. One of the important advantages of brazing is the ability to connect dissimilar materials. Brazing is also a group processing method, which allows using it for simultaneous connection any number of parts [2].

Rapidly quenched filler metals in amorphous and nanocrystalline states are often used for brazing of various materials [3]. By means of such filler metals, brazed joints are obtained in different fields such as rocket engineering that requires high-stress joints with great heat-resistance [4]. As it was mentioned above, the brazing allows to connect different materials such as ceramics and metal or metals that have properties that highly differs to each other. For example, these filler metals can be used to get a brazed a joint between tungsten with steel and vanadium alloys [5]. Properties of such brazed joints are strongly affected by microstructure and composition of brazed seam. In this field, it is important to choose carefully the composition of filler metal. Many tasks of creation joints between different materials in modern technologies that still do not have a solution. Due to it is also important to search for new alloys can be used as filler metals.

During the development of micro- and macro-alloying technologies, the first works on the creation and comprehensive study of new so-called multicomponent high-entropy alloys (HEA), including up to 5-6 basic elements, each in high concentration (from 5 at.% to 35 at.%), appeared. Alloys of the FeCrNiMnCo system were studied as the first candidates for such materials, since such alloys form a simple FCC structure and have good mechanical properties [6].

These materials, along with the typical characteristics of metal alloys, have unique and unusual properties inherent, for example, cermets: high hardness and resistance to softening at high temperatures, dispersion hardening, positive temperature hardening coefficient, high wear resistance, corrosion resistance and a number of other properties [7], [8].

It is assumed that HEAs are attractive for using in various fields, and there are studies in which they were used as filler metal [9], [10]. HEA can be used for brazing heat-resistant alloys, refractory metals, ceramics, due to their high melting point and good mechanical properties. HEA with a solid solution or eutectic structure can be obtained for brazing, having a good balance of hardness and ductility, with a melting point in the same range as the currently used filler metals, but without adding components that greatly reduce the melting temperature.

A non-equilibrium structure forms in HEAs due to the following effects present in such alloys: high mixing entropy effect; sluggish diffusion kinetics; severe lattice distortion; the cocktail effect that is the dependence of the material properties on its constituent elements and the interaction between these elements [10].

This study presents the results of researching high-entropy alloys of the Ni-Nb-Co-Fe-Cr system in the as-cast state and after rapid quenching from the melt. The results of experiments on obtaining a brazed joint of aluminum-oxide ceramics with using the filler metal of this system and the results of a studying this joint are presented.

2. Materials and Experimental Methods

The high entropy alloy $Nb_{0.73}CoCrFeNi_{2.1}$ was obtained from components with a purity 99.95 wt.% or more, with using pure metals Cr, Fe, Co, and Ni and NiNb ligature (52 wt.% Nb, 48 wt.% Ni). The as-cast ingot was remelted five times to achieve a homogeneous chemical composition.

One part of the ingot was used to obtain samples for further studying of the structure and properties, and the other to obtain rapidly quenched ribbon. The alloy $Nb_{0,93}Co_{1,16}Cr_{1,05}Fe_{1}Ni_{2,11}$ was melted in a similar manner. Obtaining the rapid quenched ribbon was carried out on the special installation "Crystal-702". In this installation, melt cooling rates to $10^5 – 10^6$ K/s are achieved. As a result, a rapidly quenched ribbon with a width of about 10 mm and a thickness of about 30 μm was obtained. Compositions of obtained HEAs are presented in the Table 1.

Table 1. Compositions of HEAs (values are shown in wt.%).

HEA	Nb	Co	Cr	Fe	Ni
$Nb_{0.73}CoCrFeNi_{2.1}$	19.0	16.5	14.5	15.6	34.4
$Nb_{0,93}Co_{1,16}Cr_{1,05}Fe_{1}Ni_{2,11}$	22.2	17.6	14.0	14.4	31.8

To determine the temperature range of melting and crystallization of the studied alloys, differential thermal analysis was carried out using differential thermal analysis (DTA). Samples of an ingot and a rapidly quenched ribbon weighing about 100 mg were prepared for measurement. The measurements were carried out at a heating rate of 10 °C/min.

The microhardness of the samples was measured on a microhardness tester model HVS-1000 under a load of 100 g for 30 s.

X-ray phase analysis was carried out on a DRON-3.0 diffractometer with the Bragg-Brentano focusing scheme using K_α characteristic radiation of copper anode. The angular velocity of the detector is 1 degree per minute.

Annealing and brazing of the samples were carried out in a vacuum furnace with resistive heating. The design of the furnace allows to achieve vacuum of $1.3 \cdot 10^{-3}$ Pa at temperatures up to 1600 °C. Annealing of the sample of rapidly quenched ribbon was carried out for 8 hours at a temperature of

1000 °C in a vacuum for homogenization of the composition and elimination of the dendritic structure formed during rapid quenching.

The samples of an aluminum-oxide ceramics XC22 were prepared for brazing. This XC22 ceramic has the next composition: 94.4 wt.% Al_2O_3; 2.76 wt.% SiO_2; 0.49 wt.% Cr_2O_3; 2.35 wt.% MnO. These samples are 1 mm thick plates with an area of about 1 cm^2. During the brazing the samples were fixed in a clamp. The brazing was carried out according to the mode, which includes heating to 1300 °C, holding for 20 minutes and step cooling.

3. Results and discussion

3.1. Calculation

The following quantities, that characterize the behavior of atoms, were calculated for HEA $Nb_{0.73}CoCrFeNi_{2.1}$: the atomic size difference δ, the mixing enthalpy ΔH_{mix} and the mixing entropy ΔS_{mix}. These quantities are $\delta = 4.8$; $\Delta H_{mix} = -12.2$ kJ/mol; $\Delta S_{mix} = 12.8$ J/(K·mol).

It is shown [11], [12] that solid solution is formed if these values are in the ranges: $0 \leq \delta \leq 8.5$; $-22 \leq \Delta H_{mix} \leq 7$ kJ/mol; $11 \leq \Delta S_{mix} \leq 19.5$ J/(K·mol). Amorphous alloys are formed when these parameters are in the following ranges: $\delta \geq 9$; $-35 \leq \Delta H_{mix} \leq -8.5$ kJ/mol; $7 \leq \Delta S_{mix} \leq 14$ J/(K·mol). However, this range may be wider if the alloy contains components with a high tendency to form an amorphous phase. With other values of the parameters, in most cases, the alloy structure contains some kind of intermetallic phase.

Thus, the calculated parameters δ, ΔH_{mix} и ΔS_{mix} predict the formation of a solid solution in the $Nb_{0.73}CoCrFeNi_{2.1}$ HEA and the impossibility of achieving an amorphous state. However, the experimental results obtained by observing the microstructure of the alloy show its pre-eutectic structure, consisting of two structural components: a solid solution and a eutectic. It will be shown below that an intermetallic compound is also in the alloy structure, which was not predicted by the calculation. In addition, the amorphous state was not achieved in the rapidly quenched ribbon.

3.2. Microstructure

The microstructure of cast ingot and rapidly quenched ribbon of HEA $Nb_{0.73}CoCrFeNi_{2.1}$ was studied. Figure 1 shows the microstructure of cast ingot. It has a pre-eutectic structure consisting of two structural components: solid solution and eutectic. To clarify the chemical composition, an EDX analysis was performed using SEM, the result of which is shown in Figure 2 and in Table 2.

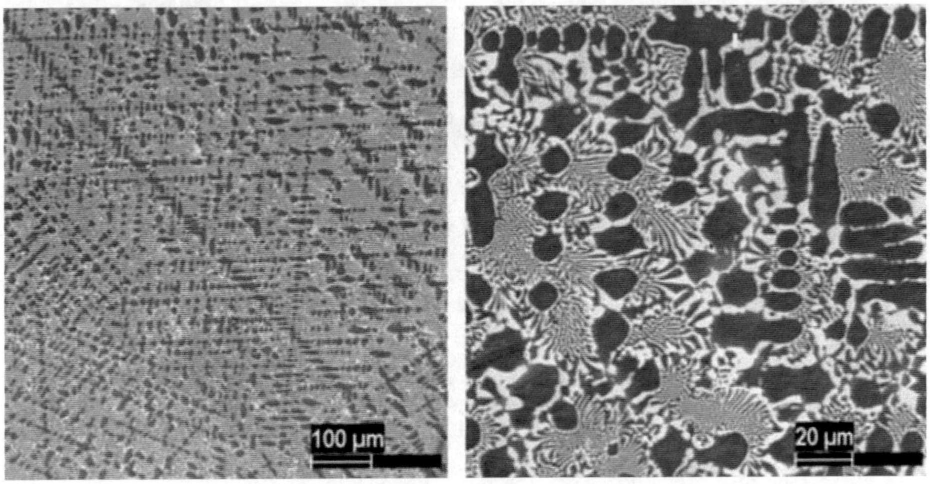

Figure 1. Microstructure of the cast ingot of the HEA $Nb_{0.73}CoCrFeNi_{2.1}$.

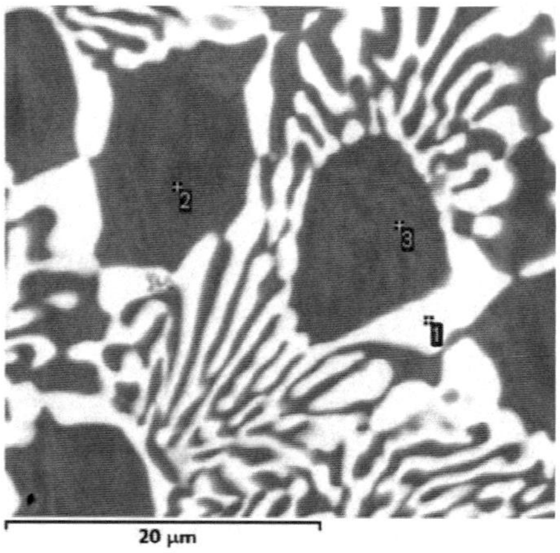

Figure 2. Microstructure of the cast ingot of the HEA $Nb_{0.73}CoCrFeNi_{2.1}$; the chemical compositions of areas 1 – 3 are shown in Table 1.

Table 2. X-ray microanalysis of the cast ingot of HEA $Nb_{0.73}CoCrFeNi_{2.1}$ (values are shown in wt.%).

Area	Cr	Fe	Co	Ni	Nb
1	12.5	12.9	19.2	30.8	24.7
2 and 3	21.3	20.0	18.2	35.8	4.7

Figure 3 shows the microstructure of the rapidly quenched ribbon of the same HEA. It has the thickness about 30 µm. In general, the microstructure of the ribbon has the same character as the microstructure of the cast ingot. Dendritic growth mainly occurred in the direction of heat dissipation. Near the contact surface of the ribbon the highest cooling rate was achieved that caused formation of an ultrafine structure in this area.

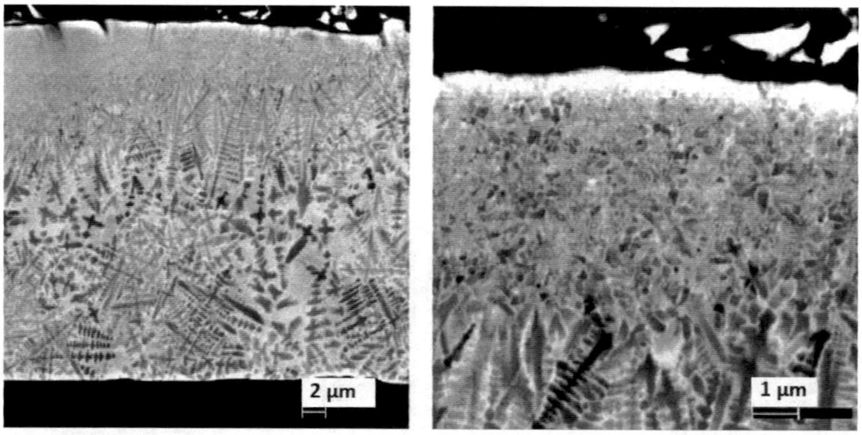

Figure 3. Microstructure of the rapidly quenched ribbon of the HEA $Nb_{0.73}CoCrFeNi_{2.1}$.

3.3. X-ray diffraction phase analysis

To determine the phase composition of the $Nb_{0.73}CoCrFeNi_{2.1}$ HEA, an X-ray diffraction analysis of the annealed rapidly quenched ribbon was carried out (Figure 4). By comparing the results with the results of the analysis of the chemical composition carried out using SEM, it became possible to conclude that the alloy contains two phases: a solid solution based on Ni with Co, Fe and Cr dissolved in it; $NbNi_3$ intermetallic compound with FCC lattice.

A similar analysis of the alloy $Nb_{0.93}Co_{1.16}Cr_{1.05}Fe_1Ni_{2.11}$ showed that it contains the same two phases.

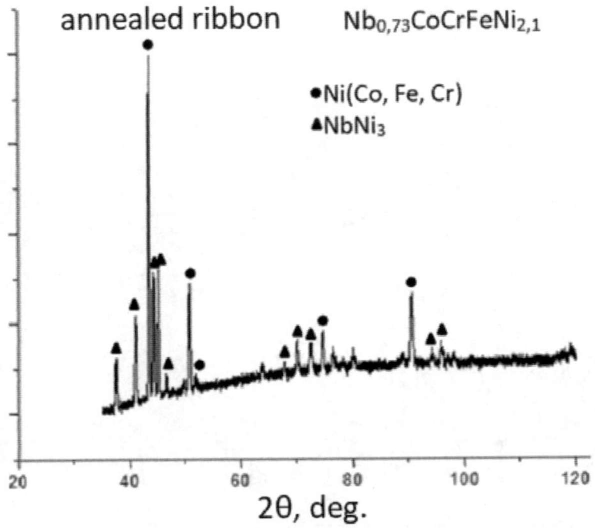

Figure 4. XRD-pattern of the annealed rapidly quenched ribbon of the HEA $Nb_{0.73}CoCrFeNi_{2.1}$.

3.4. Differential thermal analysis

Thermal curves obtained as a result of DTA of samples of the HEAs $Nb_{0.73}CoCrFeNi_{2.1}$ and $Nb_{0.93}Co_{1.16}Cr_{1.05}Fe_1Ni_{2.11}$ are shown in Figures 5 and 6, respectively. With using these results, liquidus and solidus temperatures of HEAs were determined:

- $Nb_{0.73}CoCrFeNi_{2.1}$: $T_L \sim 1260$ °C, $T_S \sim 1180$ °C;
- $Nb_{0.93}Co_{1.16}Cr_{1.05}Fe_1Ni_{2.11}$: $T_L \sim 1240$ °C, $T_S \sim 1180$ °C.

For both alloys, there is no significant difference between the rapidly quenched tape and the cast ingot. Considering this, the experiment with HEA $Nb_{0.93}Co_{1.16}Cr_{1.05}Fe_1Ni_{2.11}$ was only managed for as-cast ingot. After crystallization with further cooling down to the room temperature, no curve peaks are observed, which indicates the absence of polymorphic transformations of the studied alloys, since no processes associated with the release or absorption of a heat have been recorded.

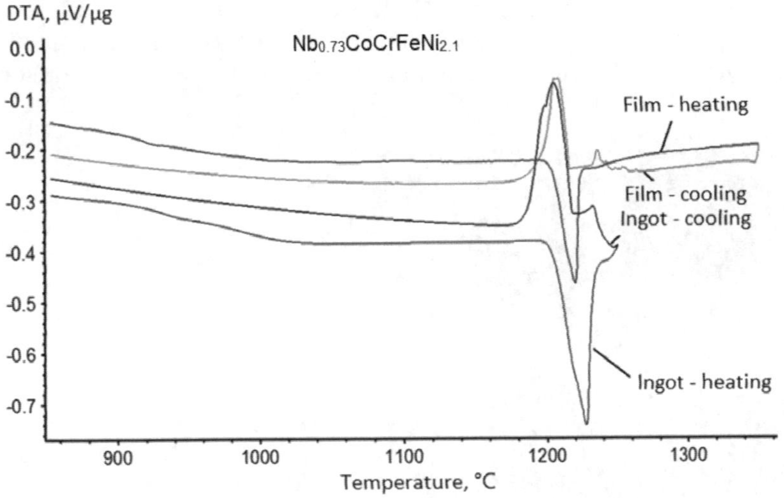

Figure 5. DTA curve for the samples of rapidly quenched ribbon and cost ingot of the HEA $Nb_{0.73}CoCrFeNi_{2.1}$.

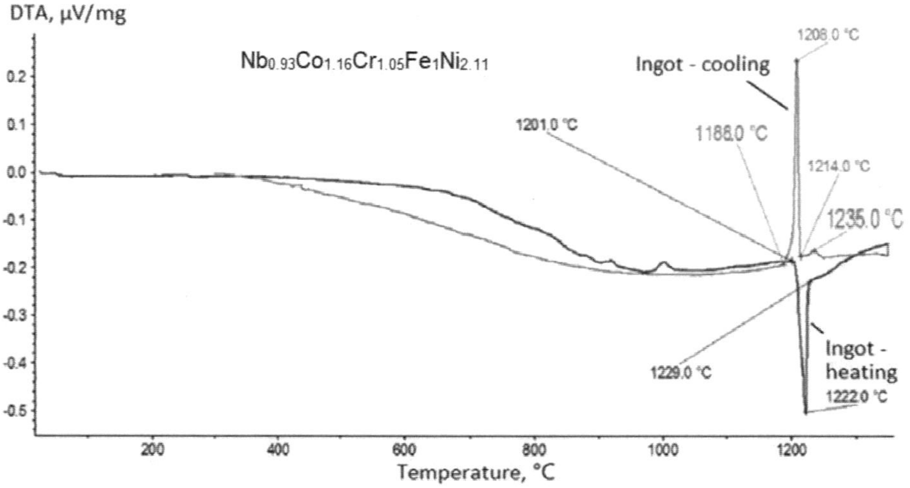

Figure 6. DTA curve for the sample of cost ingot of the HEA $Nb_{0.93}Co_{1.16}Cr_{1.05}Fe_1Ni_{2.11}$.

The HEA with optimized composition $Nb_{0.93}Co_{1.16}Cr_{1.05}Fe_1Ni_{2.11}$ contains a smaller amount of the solid solution primarily releasing during crystallization than the $Nb_{0.73}CoCrFeNi_{2.1}$ alloy. Therefore, one can conclude that the alloy $Nb_{0.93}Co_{1.16}Cr_{1.05}Fe_1Ni_{2.11}$ is closer to the eutectic composition than $Nb_{0.73}CoCrFeNi_{2.1}$. This explains its lower liquidus temperature that approximately equals 1240 °C. Based on the experiments, one can say that the crystallization of the studied alloys proceeds as follows: at a liquidus temperature of $T_L \sim 1240 - 1260$ °C, precipitation of crystals of a solid solution based on Ni with Co, Fe and Cr dissolved in it begins; as the temperature decreases a number of precipitates grows and when the solidus temperature $T_S \sim 1180$ °C is reached, a eutectic forms consisting of the same solid solution and $NbNi_3$ intermetallic compound.

3.5. Brazing of the aluminum-oxide ceramics

The rapidly quenched ribbon of HEA $Nb_{0.93}Co_{1.16}Cr_{1.05}Fe_1Ni_{2.11}$ was used as a filler metal to obtain a brazed joint of aluminum-oxide ceramics 22XC. The chosen mode with step cooling provides the minimum amount of residual thermal stresses in the brazed seam, which is important for maintaining the integrity of the joint. With the right cooling mode, cracks in the joint can be eliminated. Figure 7 shows the microstructure of the brazed joint.

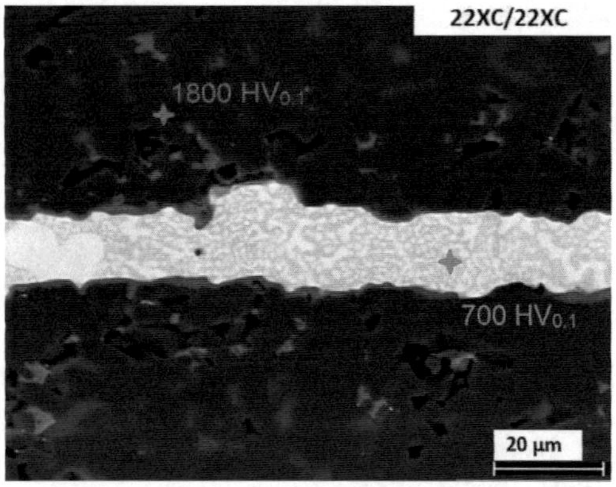

Figure 7. Microstructure of the brazed joint 22XC/22XC with microhardness data, filler metal is HEA $Nb_{0.93}Co_{1.16}Cr_{1.05}Fe_1Ni_{2.11}$, brazing temperature is 1300 °C for 20 min with step cooling.

An analysis of the microstructure shows that the original composition of the filler metal was saved in the brazed seam. The seam has a pre-eutectic structure and consists of a solid solution based on Ni with Co, Fe and Cr dissolved in it and a eutectic (solid solution + $NbNi_3$ intermetallic compound). The formation of a connecting oxide layer is clearly visible (area 2, Figure 8). Presumably, the oxide layer mainly consists of oxides of aluminum, chromium and manganese, as well as niobium oxide.

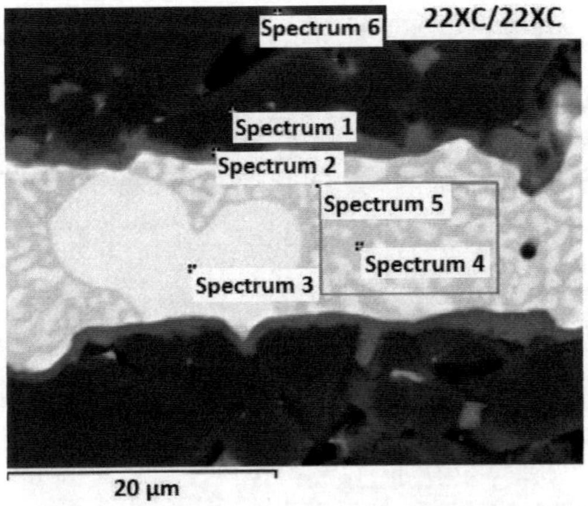

Figure 8. Microstructure of the brazed joint 22XC/22XC, filler metal is HEA $Nb_{0.93}Co_{1.16}Cr_{1.05}Fe_1Ni_{2.11}$, brazing temperature is 1300 °C for 20 min with step cooling. The results of chemical analysis in areas 1 – 6 is shown in Table 3.

Table 3. X-ray microanalysis of the brazed seam 22XC/22XC, areas 1 – 6 are shown at Fig. 8 (values are shown in wt.%).

Area	O	Mg	Al	Si	Cr	Mn	Fe	Co	Ni	Nb
1	46.9	0.0	52.3	0.0	0.4	0.0	0.0	0.0	0.4	0.0
2	33.2	0.4	**20.5**	0.3	**13.9**	**16.2**	2.1	3.2	4.6	5.6
3	0.0	0.0	0.0	1.7	6.9	0.9	11.9	18.7	24.7	35.2
4	0.0	0.0	0.0	0,0	12.8	1.7	19.9	18.8	39.5	7.3
5	0.0	0.0	0.0	0.8	10.3	1.3	16.3	18.7	34.8	17.8
6	29.3	0.0	35.3	1.4	14.8	19.2	0.0	0.0	0.0	0.0

3.6. Microhardness of the brazed joint

The microhardness of the brazed joint 22XC/22XC and the microhardness of the cast ingot are given in Table 4. It is noticeable that the microhardness of the brazed seam is slightly higher than the initial cast ingot of the HEA $Nb_{0.73}CoCrFeNi_{2.1}$.

Table 4. Microhardness of the brazed joint 22XC/22XC, the ceramic and the brazed seam.

Sample	Microhardness, $HV_{0,1}$	Error, %
Cast ingot HEA	520	12
Ceramic 22XC	1800	12
Brazed seam 22XC/22XC	700	14

4. Conclusions

Based on the research of HEAs of the Ni-Nb-Co-Fe-Cr system, the following conclusions can be made:

The following values were calculated for the HEA $Nb_{0.73}CoCrFeNi_{2.1}$: the atomic size difference are $\delta = 4.8$; the mixing enthalpy $\Delta H_{mix} = -12.2$ kJ/mol and the mixing entropy $\Delta S_{mix} = 12.8$ J/(K·mol).

The liquidus and solidus temperatures of HEAs were determined:

- $Nb_{0.73}CoCrFeNi_{2.1}$: $T_L \sim 1260$ °C, $T_S \sim 1180$ °C;
- $Nb_{0.93}Co_{1.16}Cr_{1.05}Fe_1Ni_{2.11}$: $T_L \sim 1240$ °C, $T_S \sim 1180$ °C.

The X-ray diffraction analysis results show that alloys $Nb_{0.73}CoCrFeNi_{2.1}$ and $Nb_{0.93}Co_{1.16}Cr_{1.05}Fe_1Ni_{2.11}$ contain two phases: a solid solution based on Ni with Co, Fe and Cr dissolved in it and $NbNi_3$ intermetallic compound with FCC lattice. The alloys have a hypoeutectic structure consisting of a solid solution based on Ni and a eutectic (solid solution + $NbNi_3$ intermetallic compound).

Based on DTA results the brazing mode for samples of aluminum-oxide ceramics 22XC is chosen: filler metal is HEA $Nb_{0.93}Co_{1.16}Cr_{1.05}Fe_1Ni_{2.11}$, brazing temperature is 1300 °C for 20 min with step cooling. The brazed seam 22XC/22XC has a hypoeutectic structure consisting of the solid solution based on Ni with Co, Fe and Cr dissolved in it and the eutectic (solid solution + $NbNi_3$ intermetallic compound). Presumably, the oxide layer mainly consists of oxides of aluminum, chromium and manganese, as well as niobium oxide.

The brazed seam 22XC/22XC has a microhardness 700 $HV_{0,1}$ that is slightly higher than the microhardness of the initial cast ingot of the HEA $Nb_{0.73}CoCrFeNi_{2.1}$.

As a result, it was shown that HEA $Nb_{0.93}Co_{1.16}Cr_{1.05}Fe_1Ni_{2.11}$ can be used for brazing the aluminum oxide ceramic in which the brazed seam is qualitative and has no defects.

Acknowledgements
This work was supported by Competitiveness Growth Program of the Federal Autonomous Educational Institution of Higher Education National Research Nuclear University MEPhI (Moscow Engineering Physics Institute).

References

[1] Nikolaev G, Petrunin I and Lotsmanov S 1973 *Brazing of metals* (Moscow: Metalurgia) p 275 (in Russian)
[2] Bhattacharjee T *et al* 2016 Ultrafine-Grained AlCoCrFeNi 2.1 Eutectic High-Entropy Alloy *Mater. Res. Lett.* **3** 174–9
[3] Fedotov V N, Suchkov A N, Kalin B A, Ivannikov A A and Sevryukov O N 2014 Stemet solders for brazing of modern technology materials *Tsvetnye Met.* **1** 10 (in Russian)
[4] Ivannikov A A, Sevryukov O N, Penyaz M A and Popov N S 2018 Development of nickel-based filler metal for producing high-strength joints in critical products from heat-resistant materials *Non-ferrous Met.* **45** 43–9
[5] Bachurina D, Suchkov A N, Kalin B A, Sevryukov O N, Fedotov I V, Dzhumaev P S, Ivannikov A A, Leont'eva-Smirnova M and Mozhanov E 2018 Joining of tungsten with low-activation ferritic–martensitic steel and vanadium alloys for demo reactor *Nucl. Mater. Energy* **15** 135–42
[6] Murty B S, Yeh J W and Ranganathan S 2014 *High-Entropy Alloys* (Elsevier) p 218
[7] Ivchenko M V, Pushin V G and Wanderka N 2014 High entropy equiatomic alloys AlCrFeCoNiCu: hypothesis and experimental facts *Journal of technical physic* **84** 57–69 (in Russian)
[8] Ivchenko M V *et al* 2015 Development of B2 Shape Memory Intermetallics Beyond NiAl, CoNiAl and CoNiGa *J. Alloys Compd.* **2** 5536–48
[9] Tillmann W, Wojarski L, Manka M and Timmer C 2018 Eutectic high entropy alloys – A novel class of materials for brazing applications *Proc. from Int. Brazing Solder. Conf.* April 142–8
[10] Bridges D *et al* 2018 Laser brazing of a nickel-based superalloy using a Ni-Mn-Fe-Co-Cu high entropy alloy filler metal *Mater. Lett.* **215** 11–4
[11] Guo S and Liu C T 2011 Phase stability in high entropy alloys: Formation of solid-solution phase or amorphous phase *Prog. Nat. Sci. Mater. Int.* **21** 433–46
[12] Zhang Y, Zhou J, Lin J P, Chen G L and Liaw P K 2008 Solid-solution phase formation rules for multi-component alloys *Adv. Eng. Mater.* **10** 534–8

IOP Conf. Series: Materials Science and Engineering 1005 (2020) 012013 doi:10.1088/1757-899X/1005/1/012013

Investigation on the structure of beryllium composite containing nanodiamonds

E Kozlova, M Sheverdyaev, V Gorlevsky, A Semenov, A Laushkin, A Chebotarev and A Anikin

A.A. Bochvar High-Technology Scientific Research Institute For Inorganic Materials (SC «VNIINM»), Rogova street 5a, Moscow, 123098, Russia

E-mail: keleva@mail.ru

Abstract. The article presents the results of research of the structure of advanced beryllium composite materials for use in X-ray and synchrotron technology. The technique of acquiring beryllium composite containing nanodiamonds is described. The investigation on the structure was carried out using the scanning electron microscope microanalyzer Philips XL-30. The spectral analysis of composites was carried out using an energy-dispersive spectrometer with a Sapphire detector.

1. Introduction

The materials applied in X-ray optics, for instance, in speckle suppressors should meet the following requirements: 1) to be radio transparent with minimum absorption 2) to scatter X-ray 3) to have thermal stability of 600 °C or higher. It has been established that the optimal material for making speckle-suppressor devices is a composite consisting of beryllium matrix where the carbon filler is evenly distributed in order to raise the amount of structure inhomogeneities [1]. The composite material has a high scattering ability and allows reducing the degree of radiation coherence [2]. Nevertheless, the main requirement for the composite is a high uniformity of its physical and chemical features. This uniformity could be possible only by means of a uniform distribution of all components throughout the volume.

The need for even distribution of the filler is a hard technical task, as particles of the filler tend to form agglomerate and conglomerate of various shapes [3], resulting in significant negative consequences for end-user products.

2. Materials and Methods

A technique has been proposed for creating a composite material by dispersing mixing of a thermo labile compound (beryllium hydride) and ultrafine detonation nanodiamond powder in fluid with further compaction and thermal treatment. Therefore, during the heat treatment two processes are taking place – decomposition of basic beryllium hydride and establishment of a strong adhesive link between beryllium particles and nanodiamond powder.

The resulting composite has open porosity and advanced mechanical characteristics (at the same porosity) than porous beryllium obtained by traditional ways of powder metallurgy. Furthermore, the structure of the composite consists of nanosized crystallites.

Table 1 presents composition of samples with various content of nanodiamonds.

IOP Conf. Series: Materials Science and Engineering 1005 (2020) 012013 doi:10.1088/1757-899X/1005/1/012013

Table 1. Composition of samples.

№№	Sample composition			
	Beryllium		Nanodiamonds	
	wt %	vol %	wt %	vol %
1	94.56	96.88	5.44	3.12
2	76.03	85.00	23.97	15.00
3	54.70	68.33	45.30	31.67
4	45.17	59.54	54.83	40.46

The assessment of the composites structure homogeneity has been carried out using the scanning electron microscope microanalyzer Philips XL-30 equipped with an energy-dispersive spectrometer with a Sapphire detector operating under EDAX software.

3. Results and Discussion

It was shown [1] that the maximum x-ray scattering is found in composites with nanodiamonds content from 45 to 55 wt. %. An investigation of the microstructure and elemental analysis has been carried out along the structure cross-section.

Figure 1 presents composite microstructure with filler content at 54.83 wt.% sample 4 from Table 1 with an indication of layers used to asses composition elements.

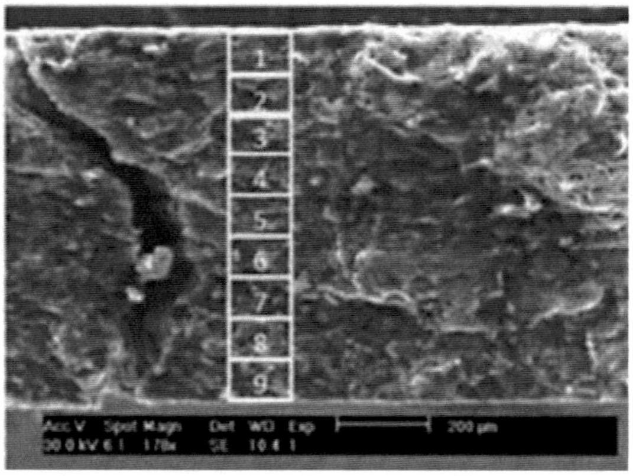

Figure 1. Structure of the composite.

Table 2 shows the results of composite analysis along the structure cross-section present in Figure 1.

Table 2. Results of the composite elemental analysis along the structure cross-section.

№№	C		O		Ca		Fe		Total	
	wt %	at %	wt %	at %	wt %	at %	wt %	at %	wt %	at %
1	74.40	79.99	24.36	19.66	0.7	0.22	0.55	0.13	100.00	100.00
2	69.36	76.32	27.61	22.80	1.71	0.56	1.33	0.31	100.00	100.00
3	72.77	78.93	25.15	20.48	1.19	0.39	0.90	0.21	100.00	100.00
4	75.10	80.77	23.24	18.76	0.94	0.30	0.72	0.17	100.00	100.00
5	74.20	79.99	24.15	19.54	0.89	0.29	0.76	0.18	100.00	100.00
6	80.74	85.57	17.52	13.94	0.99	0.31	0.75	0.17	100.00	100.00
7	72.44	78.32	26.24	21.30	0.78	0.25	0.54	0.13	100.00	100.00
8	75.84	81.15	23.11	18.56	0.58	0.19	0.47	0.11	100.00	100.00
9	80.60	85.52	17.55	13.97	0.96	0.31	0.89	0.20	100.00	100.00

Basing on the data from Table 2, it can be observed that the content of carbon particles at various layers along the structure cross-section varies from 69 to 81 wt. %, which indicates a relatively uniform distribution of nanodiamond particles in the beryllium matrix. Oxygen occurrence can be ascribed to oxide film formation at the composite surface, whereas impurities of calcium and ferrum stem from the fact that the said impurities were imported when performing the operations of composite making.

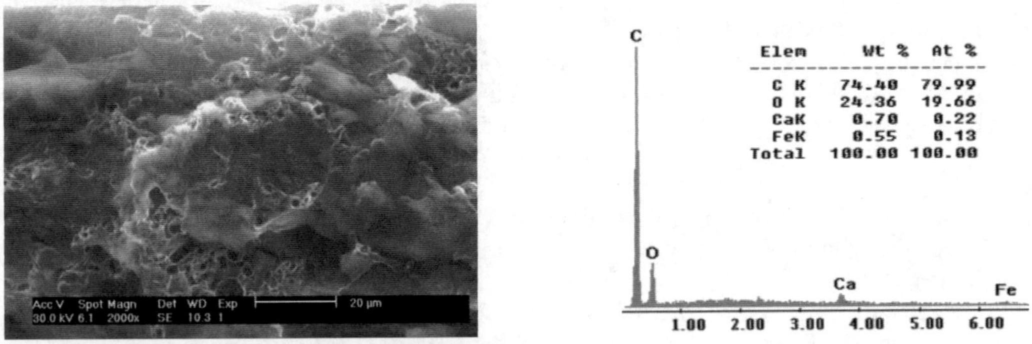

Figure 2. Microstructure and elemental analysis of the first layer composite.

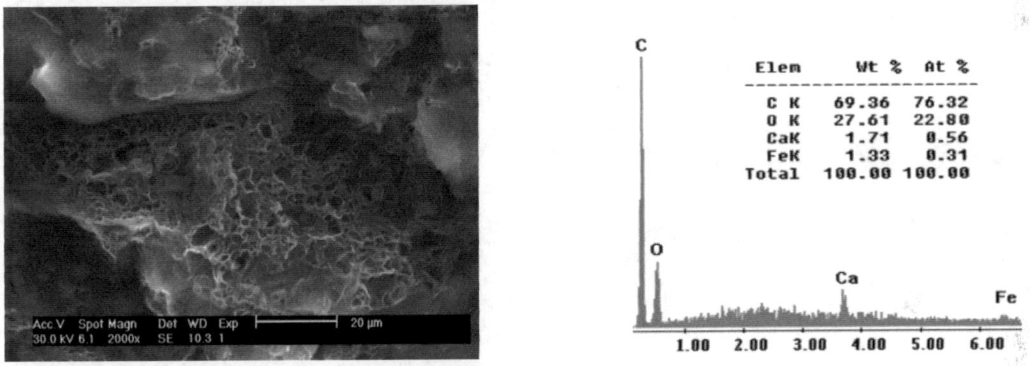

Figure 3. Microstructure and elemental analysis of the second layer composite.

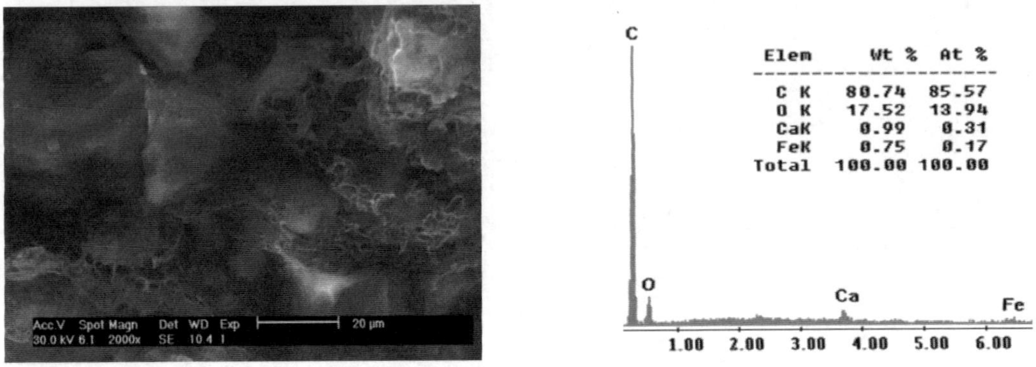

Figure 4. Microstructure and elemental analysis of the sixth layer composite.

Figures 2-4 demonstrate SEM of several layers. It can be observed that the composite structure with distributed carbon filler is represented by highly-porous microcellular beryllium matrix as the same structures as described above [4]. Carbon inclusions are spread out in the composite structure particularly as agglomerations distributed at the porous beryllium matrix, whereas the presence in the structure of inhomogeneities with an average diameter of 4-5 nm corresponding to single nanodiamond particles is confirmed by small-angle X-ray scattering [1].

IOP Conf. Series: Materials Science and Engineering 1005 (2020) 012013 doi:10.1088/1757-899X/1005/1/012013

4. Conclusion

A technique has been proposed for creating a composite material by dispersing mixing of a thermo labile compound (beryllium hydride) and ultrafine detonation nanodiamond powder in fluid with further compaction and thermal treatment.

Composites with the content of nanodiamond filler from 5 to 55 wt. % have been produced.

The investigation of elemental analysis of obtained composite demonstrated that carbon particles at various layers along the structure cross-section fluctuate from 69 to 81 wt. %, which indicates a relatively uniform distribution of the filler in the composite.

The composite structure is represented by highly-porous microcellular beryllium matrix with a carbon filler distributed in it. Moreover, carbon inclusions are allocated in the structure mostly as agglomerations distributed at the porous beryllium matrix.

References

[1] Kozlova E V, Gorlevsky V V, Semenov A A, Volkov V V, Kozlova A S, Lyatun I I, Ershov P A and Snigirev A A 2018 *KnE Materials Science* **4 (1)** 44–50

[2] Goikhman A, Lyatun I, Ershov P, Snigireva I, Wojda P, Gorlevsky V, Semenov A, Sheverdyaev M, Koletskiy V and Snigirev A 2015 *J.Synchhrotron Rad* **22** 796–800

[3] Vul' A Y and Shenderova O A 2014 *Detonation Nanodiamonds: Science and Applications* (Boca Raton: CRC Press) p 320

[4] Kostylev F A, Gorlevsky V V, Senin M D, Golikov I V, Kondrat'ev M V, Starshina V G and Chubakova T A 1995 *Neorganicheskie materialy* **31 (4)** 479–482 (in Russian)

IOP Conf. Series: Materials Science and Engineering 1005 (2020) 012014 doi:10.1088/1757-899X/1005/1/012014

Movement of discrete materials under the influence of a pulsed electromagnetic field

V Mironov, V Jakushevich and Y Usherenko

Riga Technical University , Civil Engineering Department, Powder Materials Research Laboratory, 1 Kalku Street, LV-1658, Riga, Latvia

E-mail: Viktors.Mironovs@rtu.lv

Abstract. The article considers the phenomenon of accelerated movement of small-sized objects in a pulsed electromagnetic field. For experimental studies capacitor-type pulsed current generators were used. The experiments were carried out on metal powders with a size of 50-200 microns and compact metal fragments with a weight of 1-50 g. The influence of mass, electrical conductivity and magnetic permeability of the particle material on the process of particle movement was established.

1. Introduction

The use of pulsed electromagnetic field (PEF) energy in technological processes has been known since the 70s. The most significant achievements were obtained by methods of magnetic-pulse welding and assembly of tubular thin-walled metal parts [1,2]. The method of magnetic-pulse processing of materials is used for punching holes in sheets and pipes [3], electromagnetic orientation of parts [4], deposition and removal of coatings [5]. Proposals for the use of the method to obtain one-piece compounds of powder parts are known [6,7].The phenomenon of pulsed movement (IP) of metal objects in a pulsed electromagnetic field is described, for example, in [8].

A relatively new version of the technology that allows simultaneous additional alloying and rebuilding of the initial metal material array is pulsed processing methods [9,10], which, in combination with the use of pulsed accelerators of metal particles, provides technological opportunities for industrial use.

In [11,12], the possibility of using PEF for moving powder materials was shown. The studies were carried out on ferromagnetic dispersed materials, in particular with the aim to assess the possibility of transporting of iron powders by an electromagnetic field.

It was found out that in this case, the efficiency of the process (mass of powder to be lifted, maximum lifting height, productivity) depends on many factors - the amplitude, duration and shape of the curve of the electric current passing through the inductor at the time of discharge. This limits the practical application of this method, since the movement of each material requires taking into account the effects of the totality of factors. In this case, movement is possible only for ferromagnetic materials. This limitation is especially noticeable when moving a mixture of powders from various materials. In this case, delamination often occurs due to the difference in the electrical and magnetic properties of the components. Thus, it is necessary to develop a method that eliminates these problems

2. Materials and Experimental Methods

Experimental studies were carried out using capacitor-type pulsed current generators (PCG) with an energy capacity of 0.3 to 2.0 KJ (figure 1). The main parameters of the PCG are the discharge current voltage U and the accumulated energy W (see Table 1).

Figure 1. Pulse current generators MIC-500 with a cylindrical inductor (a) and MIC-300 with flat inductor (b).

Table 1. Main parameters of pulse current generators.

PCG	Energy consumpt ion, W, kJ	Operating voltage, U, kV	Amplitude of discharge current, I, kA	Storage device capacity, C, microfarad	The pulse repetition frequency, f, Hz
MIC-500	0.5	0.9	8	1200	0.5
MIC-300	0.3	0.3	1	100	30.0

To measure the pulse currents, the Rogowski loop was used. A system to record and analyze experimental data was created. It included a digital storage oscilloscope of the GDS-810S type and a computer with a microphone-type sensor (figure 1a).

The movement of objects in the pulsed magnetic field was carried out directly by a pulsed electromagnetic field in a pipe placed in a cylindrical inductor (figure 1a), or on a pallet in the form of an aluminum disk using a flat multi-turn inductor (figure 1b)

At the output of the Rogowski loop, a voltage of 100 mV corresponds to 1 kA of discharge current in the inductor. The obtained oscillograms (figure 2) show that the maximum discharge current in the inductor winding is 6-12 kA.

Figure 3 shows an acoustic signal recording obtained using a microphone sensor. The time of flight of the object was estimated as a value equal to the difference between the values of the peaks of the pulses along the abscissa. With the pipe length S = 0.85 m, the average speed of the object was varied from 3 to 50 m/s.

IOP Conf. Series: Materials Science and Engineering 1005 (2020) 012014 doi:10.1088/1757-899X/1005/1/012014

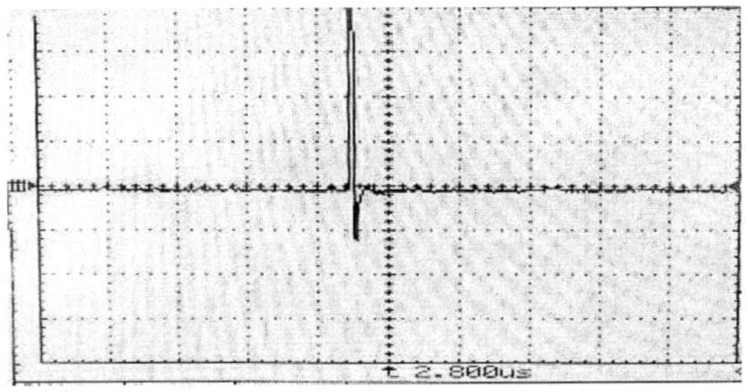

Figure 2. Oscillogram of discharge current at MIC-500.

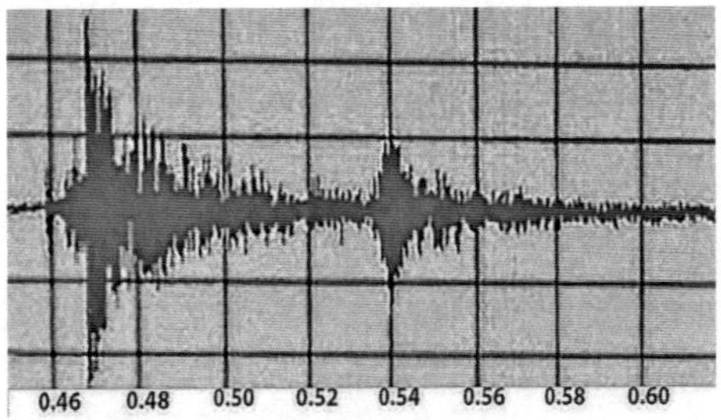

Figure 3. Acoustic diagram of the movement of the investigated object.

3. Results and Discussion

Table 2 shows the data of the pulsed movement of objects from various metal materials.

Table 2. Estimation of the speed of an object when moving in PEF.

№№	Object name	Object mass,g	Time of flight, s	Average speed, m/s	Size, mm	The magnetic permeability of the material, H/m
1	Iron powder (Fe-98%)	1.00	0.260	3.26	The average particle size – 0.08	$6.3 \cdot 10^{-3}$
2	Cast-iron shot	1.00	0.190	4.47	The average particle size – 0.14	$1.00 \cdot 10^{-5}$
3	Steel fragment	0.45	0.070	12.14	Diameter – 7.00 Thickness – 2.00	$1.26 \cdot 10^{-4}$
4	Aluminum disc No. 1	6.00	0.100	8.50	Diameter – 100.00 Thickness – 0.40	$1.25 \cdot 10^{-6}$
5	Aluminum disc No. 2	17.00	0.060	14.17	Diameter – 100.00 Thickness – 0.80	$1.25 \cdot 10^{-6}$
6	Aluminum disc No. 3	51.00	0.017	50.00	Diameter - 100.00 Thickness – 2.50	$1{,}25 \cdot 10^{-6}$

The analysis of the data shows that the presence of ferromagnetic properties of iron powder, steel and cast iron fragments provides their direct pulsed movement in an electromagnetic field. At the same time, the flight speed of a fragment of steel is higher than that of cast-iron shot and iron powder. It can be explained by the higher magnetic permeability of the steel fragment.

The movement in a pulsed magnetic field of non-magnetic aluminum disks with high electrical conductivity is of a different nature. In this case, a change in the magnetic flux in the inductor generates eddy currents in the electrically conductive object. It creates a magnetic flux counter to the main magnetic flux of the inductor. As a result of the interaction of magnetic fluxes, the aluminum disk moves in compliance with the law of inverse proportionality of speed and mass. Moreover, the greater the mass of the disk, the higher the speed of its movement.

Moving on an aluminum pallet allows you to accelerate various materials, regardless of their electrical conductivity and magnetic properties. Here, the main characteristic parameter is the mass of the object. In this case, to a certain value of the mass of the fragment, its speed increases, and then drops (figure 4).

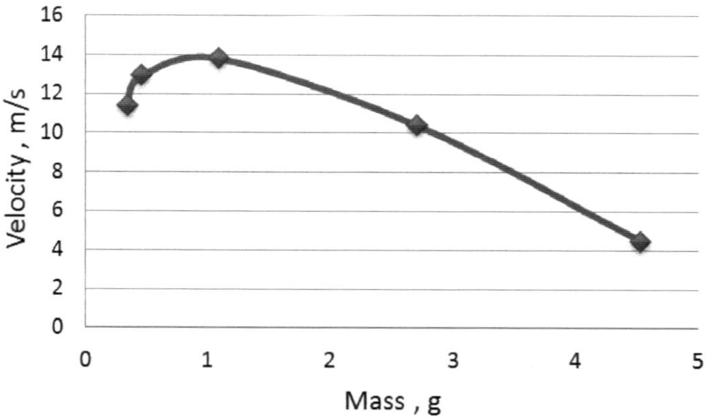

Figure 4. Change in the speed of movement in a pulsed magnetic field of steel fragments placed on an aluminum pallet.

4. Conclusion

- The presence of ferromagnetic properties in iron powders, steel and cast iron fragments makes it possible to ensure their direct pulsed movement in an electromagnetic field.
- Moving fragments on a pallet of electrically conductive material allows them to be accelerated, regardless of their magnetic properties and shows the prospects of using this method for industrial purposes

Acknowledgments

This work has been supported by the European Regional Development Fund within the Activity 1.1.1.2 “Post-doctoral Research Aid” of the Specific Aid Objective 1.1.1 “To increase the research and innovative capacity of scientific institutions of Latvia and the ability to attract external financing, investing in human resources and infrastructure” of the Operational Programme “Growth and Employment” (No. 1.1.1.2/VIAA/2/18/335).

PostDoc Latvia
NACIONĀLAIS ATTĪSTĪBAS PLĀNS 2020
EIROPAS SAVIENĪBA
Eiropas Reģionālās attīstības fonds
IEGULDĪJUMS TAVĀ NĀKOTNĒ

References

[1] Belyy I V, Fertik S M and Khimenko L T 1977 *Electromagnetic Metal Forming Handbook* (Vischa Shkola: Khar’kov) 168 pp (in Russian)

[2] Shribman V 1990 Take advantage of the new magnetic pulse welding process 2001 *Svetsaren: Welding Review* **2–3** 14–16

[3] Gluschenkov V A The main directions of development of magnetic-pulse processing *Forging and stamping production* **7** 11–13 (in Russian)

[4] Ioffe B 1972 *Electromagnetic orientation of parts* (Riga: Zinatne) 300 pp (in Russian)

[5] Dorozhkin N, Mironov V and Kot A 1985 *Electrophysical methods for producing coatings from metal powders* (Riga: Zinatne) 131 pp (in Russian)

[6] Mironovs V, Boiko I and Kolbe M 2015 The Application of the Pulsed Electromagnetic Field for Joining Powder Parts *Solid State Phenomena* **220** 864–870

[7] Mironov V, Kolbe M, Tatarinov A and Gluschenkov V 2016 Detachment of conductive coatings by pulsed electromagnetic field *Proc. 7th Int. Conference on High Speed Forming* (ICHSF-2016, Dortmund) 69–77

[8] Gafri O, Izhar A,Livshitz V and Shribman V 2006 Magnetic pulse acceleration *Proc. 2nd Int. Conference on High-Speed Forming* (Dortmund) 33–41
[9] Usherenko Y, Usherenko S and Yazdani J 2017 High-Energy Method of Transformation of Casting Metals and Alloys to the Composite Materials *Key Engineering Materials* **721** 290–294.
[10] Usherenko Y, Mironovs V, Lapkovskis V, Usherenko S and Gluschenkov V 2019 Powder particle flow acceleration methods for simulation of interaction with materials used in spacecrafts *Agronomy Research* **17(6)** 2445–2454
[11] Mironov V and Viba J 2007 Elevator for Powders *In: Proc. EuroPM-2007*, Toulouse, **2** 39–44
[12] Lapkovsky V, Mironovs V, Shishkin A and Zemchenkov V 2011 Conveying of Ferromagnetic Powder Materials by Pulsed Electromagnetic Field *Proc. EuroPM-2011 – Tools for Improving PM: Modelling & Process Control*, Barcelona 253–258

IOP Conf. Series: Materials Science and Engineering 1005 (2020) 012015 doi:10.1088/1757-899X/1005/1/012015

Regularities of instantaneous (flash) spark - plasma sintering of silicon carbide

V Kazakova, D Shornikov, A Tenishev and E Shornikova

National Research Nuclear University MEPhI (Moscow Engineering Physics Institute) Kashirskoe highway 31, Moscow,115409, Russia

E-mail: kazackovavictorija@yandex.ru

Abstract. This paper shows the possibility of producing high - density materials based on silicon carbide in a short time with the methods of electropulse consolidation and flash-sintering. During the experiments, several types of modes were considered which vary from each other by different initial data (pressure, temperature, power).While analyzing the data, a flash - sintering mode has been improved. The sintering of silicon carbide with the given mode allows producing a sample with a density equal to 97% of its theoretical density.

1. Introduction

Nowadays, ceramics previously used as building products and household goods has changed into a multifunctional material and gained widespread application in such areas as electrical engineering, electronics, and mechanical engineering, where strength and resistance to brittle fracture are specially required [1–2]. The remarkable thing is that the strength of ceramics has increased by 3-5 times over the past 100 years. Products made of high-performance ceramic materials have taken a central place in many technical fields.

It is found that at present ceramics, mainly containing SiC, Al_2O_3, ZrO_2, accounts for more than 20% of the world production of structural materials, with their role tending to increase in the future[2].

The use of technical ceramics has been expanding into new applications. Among those ones used as structural ceramic materials, silicon carbide (SiC) is of great interest. SiC is believed to find application in mechanical engineering at high temperatures due to its file hardness and good thermal characteristics [3–5]. Unfortunately, the synthesis of ceramics based on silicon carbide requires high temperatures, and additive-free sintering at 2150 – 2200°C does not allow producing non-porous roducts both due to low diffusion coefficients and high volatility of SiC at 2150 – 2200° C.

The technology of sintering high-quality products from silicon carbide is not yet widespread. Consequently, there is a need for searching for ways to manufacture new types of materials with a high level of physical and mechanical characteristics, but a lower sintering temperature.

2. Materials and Experimental Methods

All experiments were performed with the SiC powder. It is a powder of 6-8 microns in size with splintery particles. The SEM image of the powder is shown in Figure 1. The powder X-ray diffraction pattern indicates the presence of the Si_3N_4 phase, i.e. the powder contains a small amount of impurity.

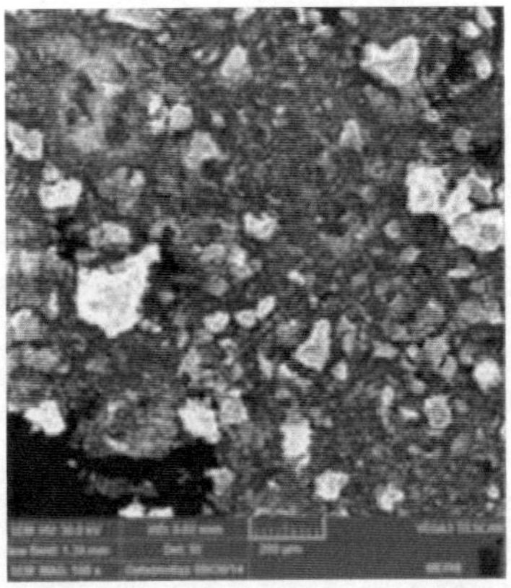

Figure 1. The SEM image of the SiC powder.

The early stage of work is SPS-sintering of SiC powder in the Labox-625 installation. The principle is to exert a combined effect of the pulsed direct current and mechanical pressure on the powdered material [6–8].

The next stage of work is the sintering of the produced samples in the hybrid SPS Labox-125 installation. The peculiarity of the hybrid SPS installation involves two-stage heating: the sample is preheated to a certain critical temperature level due to thermal radiation; then, upon reaching a certain temperature, the sample becomes conductive, and a direct electric current is passed through it, ensuring its rapid heating directly inside the sample (Figure 2) due to the Joule heat generation.

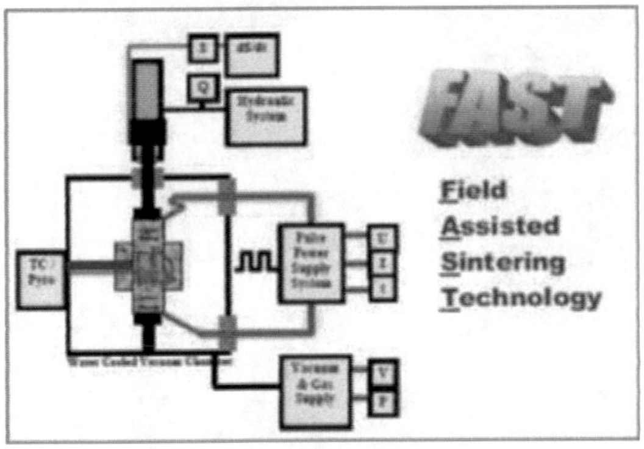

Figure 2. The principle underlying the FAST method.

The optimum mode selection for flash sintering of SiC depends largely on the adequate determination of the critical temperature value at which the spark-plasma sintering process should be launched. As a rule, this temperature lies in the transition region of the semiconductor from impurity to intrinsic conductivity.

In order to obtain information on the structure and grain size in sintered pellets, images were taken on SEM of the TESCAN firm (model VEGA3). Microhardness of the sintered samples was measured using the Vickers method. For this purpose, an FM-800 microhardness meter (Future - Tech) with automatic loading was used. The loading holding time was $\tau = 15$ s, loading mass was 300g.

3. Results and Discussion

SPS - sintering

To perform "flash sintering" experiments, cylindrical samples of silicon carbide with a relative density in the range of 70–80% were preliminarily prepared. These sample were sintered in a standard SPS installation in a graphite matrix. The sintering temperature for the sample is presented below in Figure 3.

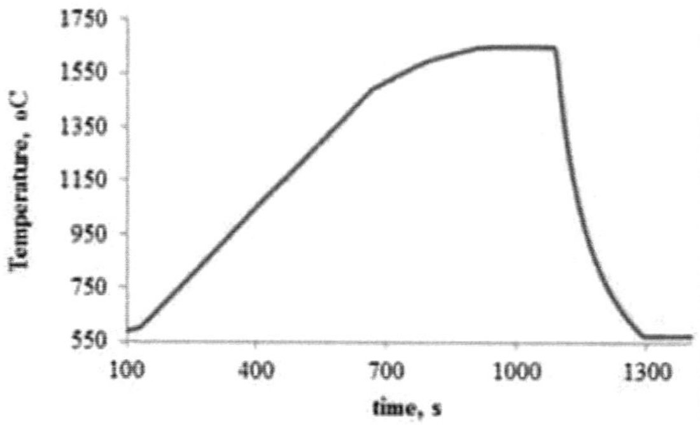

Figure 3. SPS sintering temperature regime of a coupon in the Labox-625 installation.

Heating was performed with passing a pulsed direct current at a heating rate of 100°C/min. The temperature was monitored by the outside of the graphite matrix using an optical pyrometer. The filling was loaded immediately before the heating with a constant load (at a pressure of 30 MPa), the load was removed after the heating stopped. The high-temperature holding time was 3 minutes. After completing the high-temperature holding time, both the sample and chamber were cooled. When the temperature of the sample reached 100° C, it was removed. Then the sample were air-dried before further measuring the density. Figure 4 shows the structure of the sintered sample – the relative density is 72% of its theoretical density.

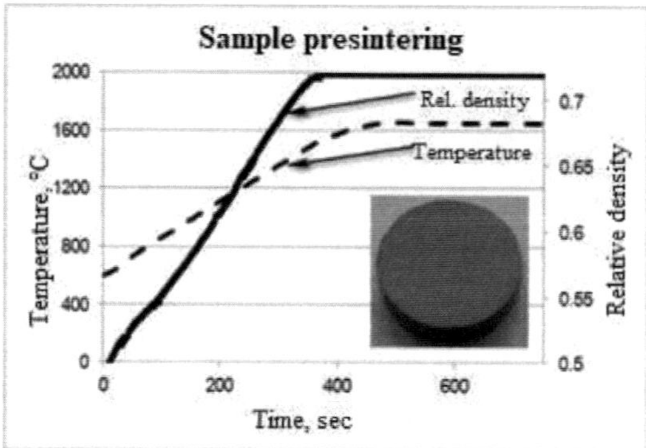

Figure 4. Sintering mode of samples before experiments on flash SPS.

Ultrarapid compaction under conditions of flash SPS is due to the effect of thermal explosion that occurs inside the powdered material by the self-accelerating Joule heating [9]. The conditions of flash sintering depend on the applied external pressure, the characteristics of the supplied power, which, while being under extreme conditions, can lead to an uncontrolled thermal explosion. The consequences of this process can be a significant heterogeneity in temperature, and, therefore, the heterogeneity of density distribution and structural parameters within the sintered volume. Under conditions of flash SPS, such inhomogeneities are equalized due to more intensive mass transport under the influence of applied pressure.

In case of most carbide systems, mass transport of the material is activated [9] as the temperature and sintering time increase. Silicon carbide is characterized by a low activity of diffusion processes [9], therefore, grain growth is insignificant during pressureless sintering (SiC grains are 2–3 times larger than particles of initial powders). During SPS sintering, mass transport of the material and grain growth are almost non-existent due to a high rate of the process.

IOP Conf. Series: Materials Science and Engineering 1005 (2020) 012015 doi:10.1088/1757-899X/1005/1/012015

Flash sintering

Based on the above assumptions, a temperature range was determined (1150 – 1400 °C), in which it is expedient to carry out flash sintering.

Figure 5 shows the sample loading into a hybrid sintering unit. A graphite punch with a diameter of 15 mm is mounted on the lower spacer, then one layer of graphite foil is laid between the punch and the coupon on each side to protect against baking-on.

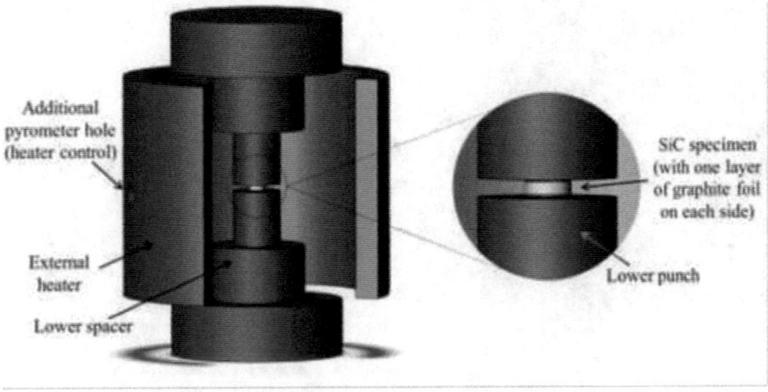

Figure 5. Equipment to carry out an experiment in the hybrid unit.

Pressure was applied to the sample at room temperature (up to 50 MPa). Then, two pyrometers were set up on the side surface of the sample, with one of them being responsible for monitoring the temperature from the external heater (hole on the left), and the other being responsible for monitoring the temperature from the SPS module (the gap in the heater is in front).

The heating mode of the sample with an external heater was set in the following way: at a rate of 30°C/min to a temperature of 1000°C, then heating to a temperature of 1150°C – 1300°C with a rate of 20°C/min. The temperature regime is shown in Figure 6.

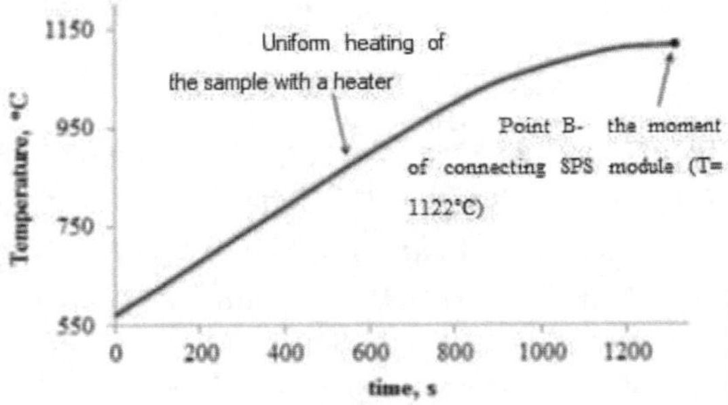

Figure 6. Temperature regime of heating with an external heater in a hybrid unit.

Nevertheless, the choice of a sintering mode depends not only on the temperature, but also on the initially applied power and pressure. Maximal parameters for SPS module: I = 1.5 kA and V = 10 V. Maximum power: P = VI = 15 kW.

An optimal mode of "instantaneous sintering" was selected.

Depending on the combination of parameters (power, pressure, temperature) set during the experiment, different data were obtained. In analyzing them, all the results can be divided into three groups.

In the first case, when the SPS module is connected, that is, when the current is supplied to the sample, the voltage reaches its maximum value of 10 V, and the current slightly increases. As shown in Figer 8, its maximum value is 45 A, hence it follows that the sample does not virtually conduct electric current, which would lead one to conclude that the resistance of the sample is still sufficiently large at this stage. Consequently, the sample conductivity is insufficient, therefore no shrinkage occurs.

Figure 7 shows large pores. The temperature jump may be due to the fact that initially a large amount of power was applied manually, which is used to heat the sample. The selected mode does not change the parameters of the sintered samples.

Figure 7. Microstructure of the sintered sample (P = 750 W).

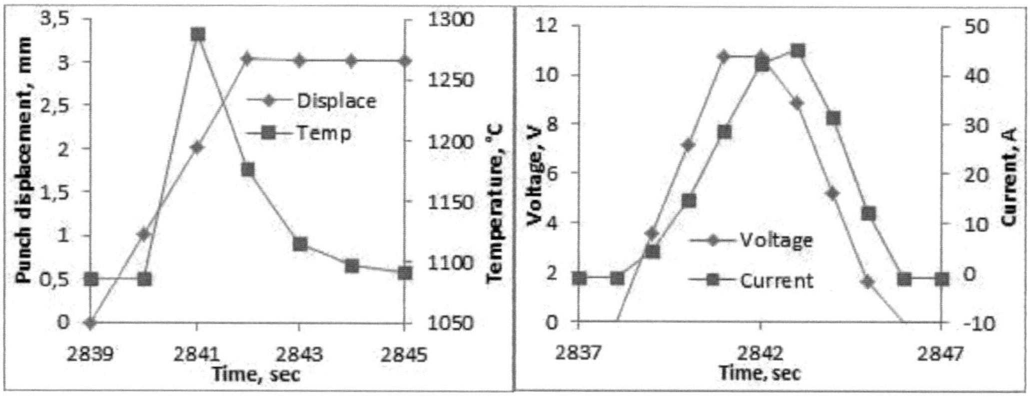

a) b)

Figure 8. *a* – The dependence of the electrode-punch displacement and temperature changes on time (P = 750 W); *b* – The dependence of current and voltage on time (P = 750 W).

In the second case, when the SPS module is connected, the sample shrinks immediately, which lasts for 2 to 3 seconds (step from 1625 to 1628 s). Figure 10 shows the electrode-punch displacement as a function of time, in accordance with the change in temperature of the sample over time. The sample is uniformly heated with an external heater until the point 1625 s. It can be seen from Figure 12 that the voltage value reaches about 8 V, a rather large current passes through the sample (about 180 A), which indicates that the sample becomes conductive. The current value is greater than in the first case, due to which there is an intense heat release and, as a consequence, an increase in the sample temperature. Sample heating in such conditions leads to “thermal explosion”. During similar experiments all samples would destroy. Figure 9 shows the sample microstructure, which is more perfect than in the previous case.

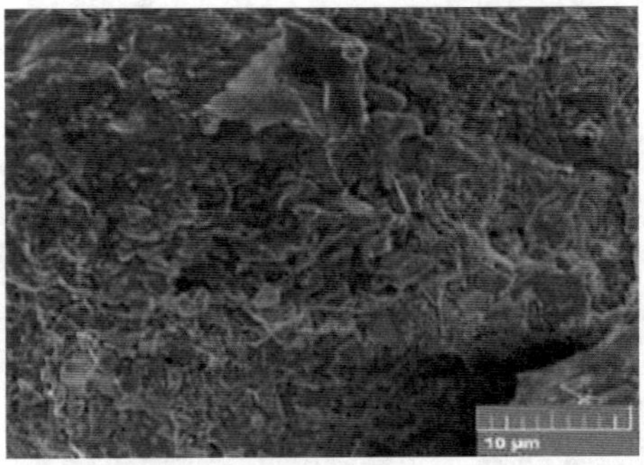

Figure 9. Microstructure of the sintered sample (P = 1200 – 1500 W).

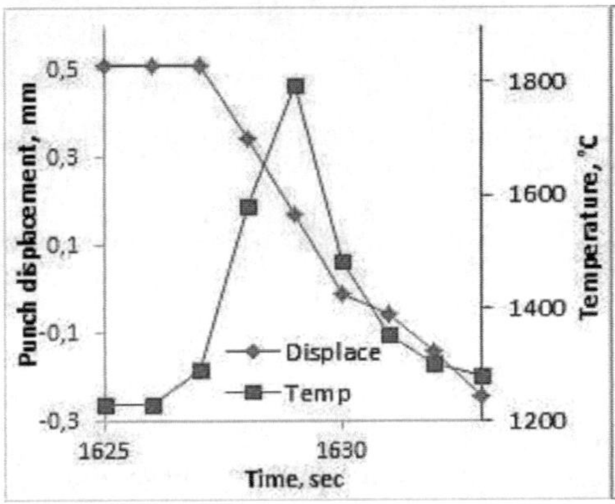

a)

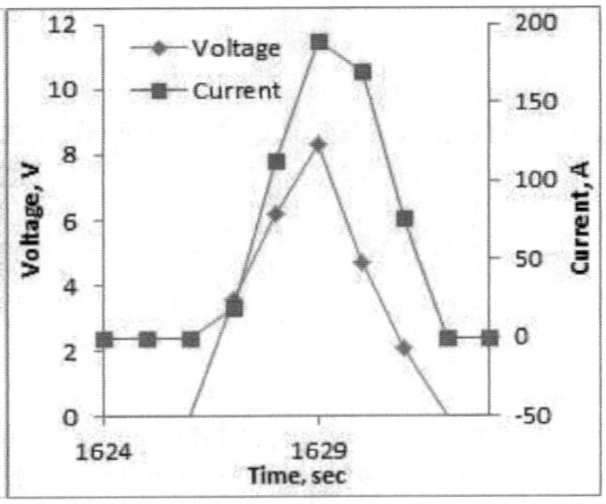

b)

Figure 10. *a* – The dependence of the electrode-punch displacement and temperature changes on time (P = 1200-1500 W); *b* – The dependence of current and voltage on time (P = 1200-1500 W).

In the third case, when the SPS module is connected at the initial moment of time, the voltage value increases to 8 V, the current value - to 95 A, and the sample shrinks during the first 2 seconds (Figure 12). The sample is heated at an earlier stage. When passing the current, the sample temperature is 1170°C, and it conducts an electric current (Figure 12). Adequate sintering mode parameters have made it possible to monitor a heating process and produce a sample with high density without destroying it. Figure 11 shows the regions with increased density. At a certain time (1357 seconds), the unit was shut down, and the sample started to cool down.

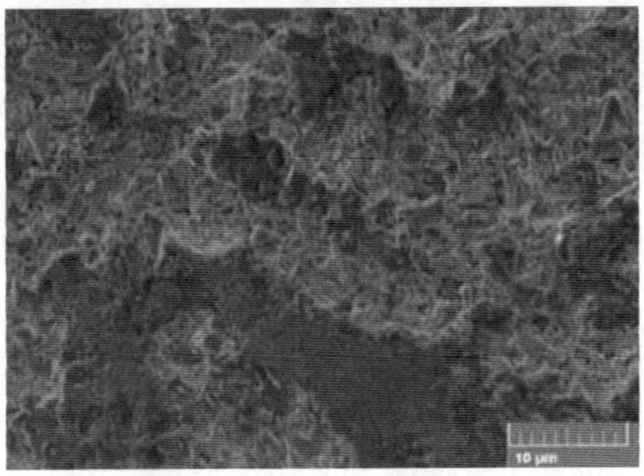

Figure 11. Microstructure of the sintered sample (P = 900 W).

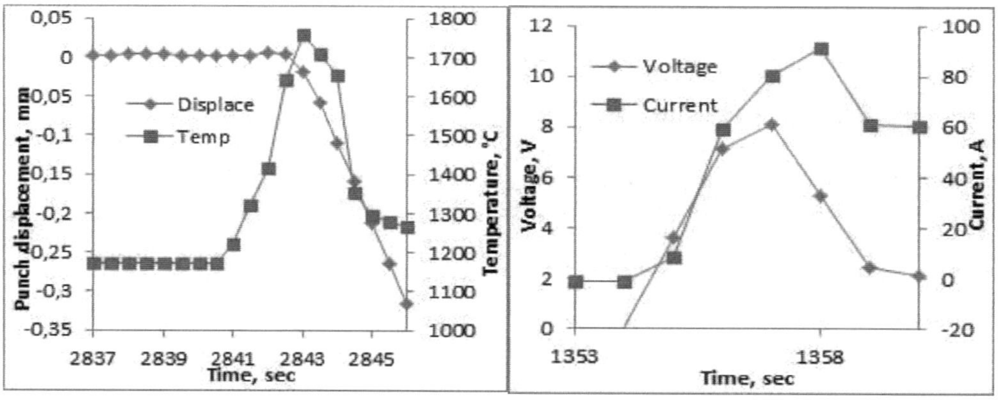

a) b)

Figure 12. *a* – The dependence of the electrode-punch displacement and temperature changes on time (P = 900W); *б* –The dependence of current and voltage on time (P = 900W).

According to the results of all experiments, a sample with a density of ρ = 97% was produced. Thus, the parameters of flash sintering were selected, which were: power – 900W, pressure – 2.36 kN, temperature of the SPS connection – 1200°C.

4. Conclusion

Based on the analysis of the research results, the following conclusions can be drawn:

The possibility of producing high density materials in a short time with a combination of methods of electropulse sintering and flash sintering, which reduces the time to produce a high-density SiC sample from several hours to several minutes.

During the experiments, several types of modes have been considered which vary from each other by different initial data (pressure, temperature, power). In analyzing the data obtained, a mode of flash sintering has been selected which can facilitate the production of a sample with a density equal to 97% from the theoretical density.

Acknowledgements

This work was supported by Competitiveness Growth Program of the Federal Autonomous Educational Institution of Higher Education National Research Nuclear University MEPhI (Moscow Engineering Physics Institute).

References

[1] Abraham T 2004 *Journal of the American Ceramic Society* **8** 23
[2] Matthews F and Rolings P 2004 *Composite materials. Mechanics and technology. Russia: Tec nosphere* 408
[3] Wang F, Yao D X, Xia Y F, Zuo K H, Xu J Q and Zeng Y P 2016 *Ceram. Int.* **42** 4526–31
[4] Wu H B, Li Y S, Yan Y J, Yin J, Liu X J, Huang Z R, Lee S H and Jiang D L 2014 *J. Eur. Ceram. Soc.* **34** 3469–78
[5] Wang F, Yin J W, Yao D X, Xia Y F, Zuo K H, Xu J Q and Zeng Y P 2016 *Mater. Sci. Eng.* A **654** 292–297
[6] Yang X, Wang X, Wang P, Hu Ke, Li Zhi and Zhang J 2017 *Journal of Alloys and Compounds* **704** 329–335
[7] Lee Y et al. 2018 *Scripta Materialia* **143** 188–190
[8] Dijen F K and Mayer E 1996 *J. Eur. Ceram. Soc.* **16** 413–420
[9] Olevsky E, Rolfing S, Maximenko A and Ilyina A 2016 *Sci. Rep. J* **6** 20

Obtaining permanent connections by electromagnetic method

S Ageev[1,2], S Nescoromniy[2] and E Strizakov[2]

[1] Merzhanov Institute of Structural Macrokinetics and Materials Science Russian Academy of Sciences (ISMAN), Academician Osipyan str., 8, Chernogolovka, Moscow Region, 142432, Russia
[2] Don State Technical University
Gagarin square 1, Rostov-on-Don, 344000, Russia

E-mail: stageev@inbox.ru

Abstract. The article deals with the technological processes of magnetic pulse crimping for the production of non-removable metal, metal-glass and metal-ceramic units. The analysis of connection of stranded wires with tips of onboard cable networks of aviation and space equipment. Assembly of metal-glass fuses and metal-ceramic knots of electrovacuum devices is carried out. The use of uniform remote electromagnetic influence on the assembled units allows for more dense packing of conductive conductors in the contacts, eliminate soldering and provide heat dissipation in the insulation units of ultrahigh frequency devices.

1. Introduction

In the production of electronic equipment, aircraft and spacecraft for composite compounds, magnetic pulse processing, which is characterized by a remote uniform and strictly dosed effect on the interface zone, is becoming increasingly common [1,2].

The most widespread are the Assembly processes using electromagnetic pressure according to the «crimper» scheme of axisymmetric composite connections of stranded cables, metal-glass and metal-ceramic assemblies [3,4].

2. Magnetic pulse crimping of stranded cables

The principle of magnetic pulse crimping pliers for sealing electric wires in the terminals or the capacity of multicore cables is as follows, see Figure 1.

The tip 3 is installed in the inductor (working tool) 2 which is connected to the pulse current generator 1. When passing through the cylindrical inductor 2 current pulse Id there is an alternating magnetic flux H, which provides the tip of the induced current flow Ii. The interaction of the induced current with the primary magnetic flux leads to the appearance of radially acting ponderomotor forces-electromagnetic pressure Pm, due to which the deformation of the tip and compaction of the internal volume of the tip is carried out. Power consumption of the equipment – 10 kJ, natural frequency of discharge current 20 kHz, amplitude value of discharge current up to 200 kA.

When the discharge of the GPC on the coils of the inductor, a current pulse flows, the magnetic field of which induces induced currents in the pusher. The interaction of induced currents in the pusher with the magnetic field of the inductor leads to the appearance of ponderomotor forces – a pulse of magnetic pressure.

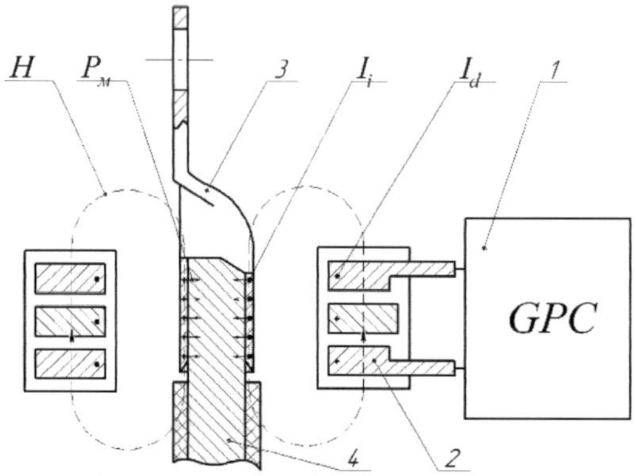

Figure 1. Schematic diagram of electromagnetic crimping for sealing electrical wires in the tips 1 – pulse current generator (GPC); 2 – inductor; 3 – tip; 4 – stranded wire; I_d – discharge current; H – magnetic flux; I_i – induced current; P_m – electro-magnetic pressure.

In contrast to the currently existing cold crimping with a special tool that creates wire contact at several points, magnetic pulse crimping provides uniform crimping and contact over the entire interface surface between the tip and the live conductors.

The use of uniform remote electromagnetic action provides a more dense packing of conductive conductors in the contacts, which allows you to enter an additional number of conductive conductors into the contacts, increasing the cross-sectional area, reducing the transient resistance, reducing losses on ohmic heating, increasing the strength of the connection, ultimately reducing the weight and increasing the reliability of on-board cable networks [3-5].

The authors conducted a study of the process of high-speed deformation of the tip material and its introduction into conductive veins. The dependences of the mechanical strength and transient resistance of the connection on the input energy of the copper tip with copper and aluminum wires are obtained, Figure 2.

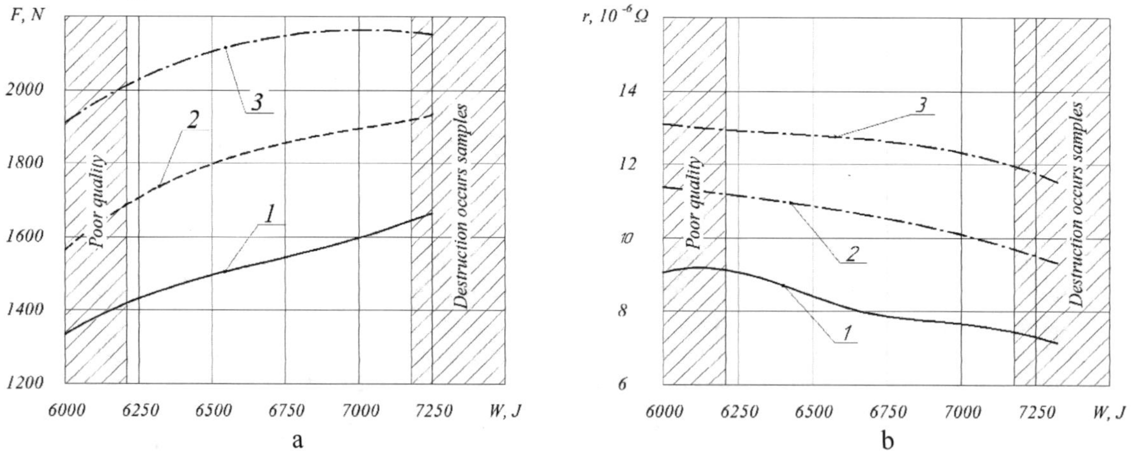

Figure 2. Dependence of mechanical strength (a) and transient resistance (b) on the pulse energy W.

Compression technology contacts and splice wires and cables by the method of magnetic-pulse processing of materials can be used not only in the manufacture of onboard cable network, but in other industries, particularly in the power sector in the creation and repair of power networks (cable and overhead lines), for any purpose, instrumentation, aviation, automotive and machine building enterprises of the defense complex, agriculture, etc.

3. Multi-pass magnetic pulse crimping of metal-glass and metal-ceramic units

In the production of products in electronic technology, instrument-making there is a need to combine heterogeneous materials, such as metal-ceramics, metal – glass.

The use of traditional methods of Assembly of multi-element compounds is ineffective and leads to irrationally high costs of materials and energy. The mechanical effect leads to the emergence of cracks or complete fracture of glass or ceramic. These disadvantages are eliminated by the use of Assembly using the energy of the electromagnetic field. It is proposed to use magnetic pulse processing.

In the process of magnetic pulse Assembly, the parts from the stack and ceramics are combined with the metal into a mechanically strong one-piece unit by plastic deformation of the enveloping conductive shell. However, intensive single-pass exposure with a high rate of deformation in this case is unacceptable. Magnetic pulse Assembly of metal-glass units (MGU) must be carried out by a series of pulses of low discharge energy (100-200 J). During the sampling of the gap between the shell and the glass frame due to induced currents, the metal is deformed and simultaneously heated. After cooling, tension is carried out in the connection without destroying the fragile base.

An example of the Assembly of a metal-glass fuse Assembly is shown in Figure 3.

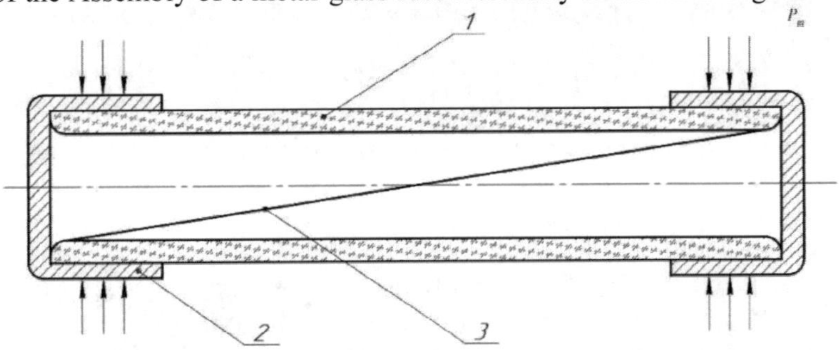

Figure 3.Magnetic pulse crimping of metal-glass fuse.

1 – glass frame (tube); 2 – fusible element; 3 – tip; P_m – electromagnetic pressure.

Figure 4 shows a composite metal-ceramic film resistor pressed by electromagnetic forces. Here is also used in the Assembly of the multiple electromagnetic loading.

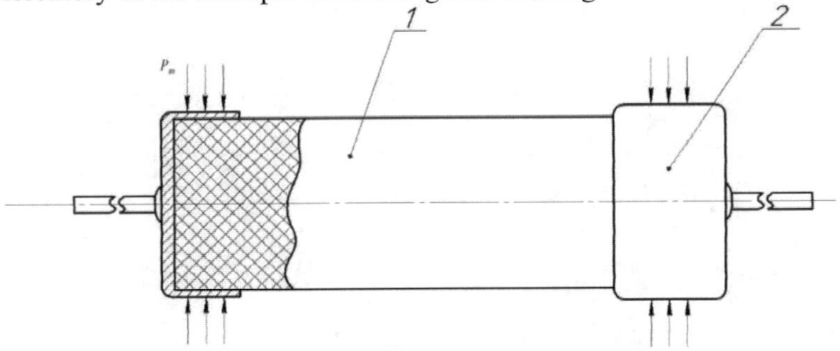

Figure 4. Scheme of electromagnetic crimping of film resistor
1 – frame; 2 – tip; P_m – electromagnetic pressure.

Composite multi-element metal-ceramic unit (MCU), shown in Figure 5, is a housing and insulation element of the microwave (SHF) electric vacuum device, which ensures the performance of structures at operating voltages of more than 10 kV and intense thermal loads.

When processing MCU mechanical methods, crimping Cams or rolling rollers in beryllium ceramics occurs foci of stress concentration, which lead to the destruction of ceramic parts, which is unacceptable in the production of microwave vacuum devices such as traveling wave lamps.

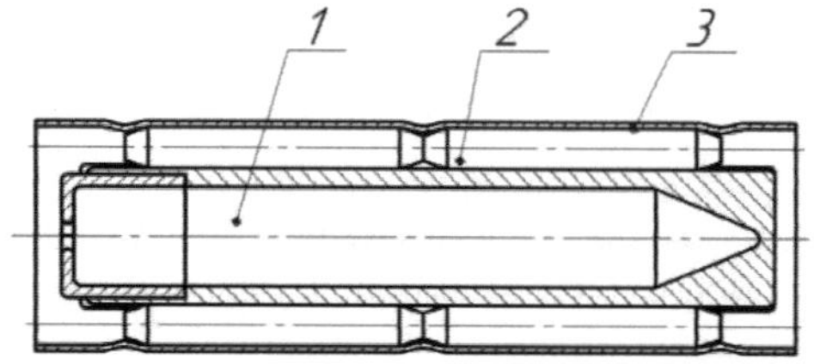

Figure 5. Composite metal-ceramic unit (MCU)
1 – base; 2 – ceramic inserts; 3 – electrically conductive shell.

It is proposed for the Assembly of multi-element MCU to use a non-contact method of processing with a strict dosage of pressure by a pulsed magnetic field. In most cases, as previously mentioned, the process is not realized in one transition, since when the necessary configuration of the metal shell is reached in the ceramic rods, microcracks are formed as a result of intensive loading.

The conditions for qualitative crimping of metal shells on ceramic bases are formulated in [6] as follows:

$$u_f>[u];\ v_c \le [v];\ s \ge [s];\ J_m \le [J_{m.max}] \quad (1)$$

These terms are controversial, as is necessary to carry out the final significant movement of the model the u_f excess of set [u] with restrictions [v] impact velocity v_c shell with ceramics, the limitations of the momentum of the first half-wave magnetic pressure $[J_{m.max}]$ and the allowable minimum thickness of the shell [s].

The critical values of collision velocities and magnetic pressure are established empirically. For different ceramic materials, their values are in the ranges: v=0.5-3 m/s, J_m = (0,1 – 2)·10 N·s/m^2.

It is established that the conditions (1) can be realized only by multi – transient processing-multiple loading with a change (increase) of the load-the magnetic pressure pulse.

The theoretical analysis of the MCU crimping process was carried out by numerical simulation on a computer of a multi-transient magnetic pulse effect by repeatedly solving a single-transient problem using the equilibrium state of the system obtained from the previous solution as the initial state.

Figure 6 shows the design scheme, the configuration of the structure and the nature of the distribution of magnetic pressure at different stages of the process of crimping MCU, photos of typical products in Figure 7.

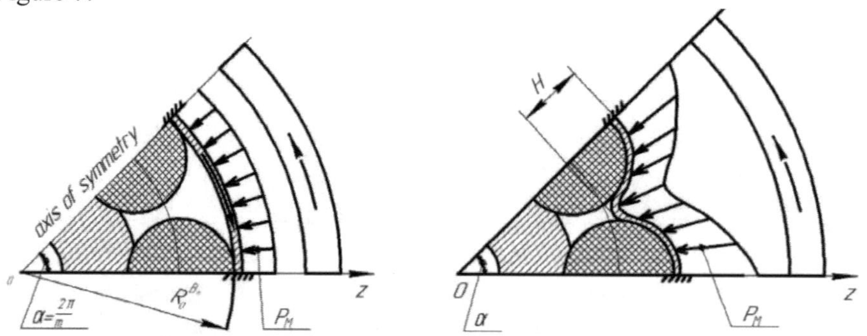

a b

Figure 6. The nature of the distribution of magnetic pressure at different stages of the process of crimping MCU: a – the initial state; b – in the final stage.

Due to the cyclic circumferential symmetry in the process of numerical studies, only the part of the product that has the form of a sector with a turning angle l = 2π/m, where m is the number of ceramic rods in the cross section of the structure, was considered. Similar MCU SHF EVD is shown in Figure 7.

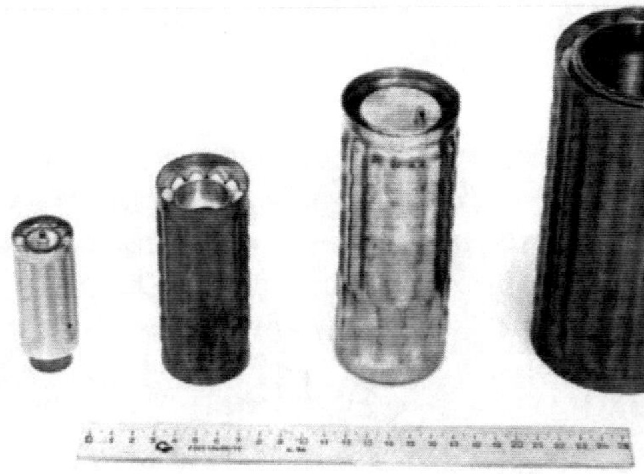

Figure 7. Typical MCU products made by magnetic pulse crimping.

Theoretical analysis and experimental studies using high-speed photoregistration of the MCU crimping process showed that the intensity of shaping in each loading pulse practically stops after the end of the first half-life of the load, which determines the role of the pulse of the first half-wave of magnetic pressure in shaping.

The analysis of loading conditions confirmed the assumption that the processing should be carried out in a series of pulses, increasing the deflection of the shell, taking into account that from transition to transition the equivalent inductance of the inductor-workpiece system increases, the magnetic pressure in the deformation zone decreases, and the deformation resistance as a result of hardening (hardening) increases. As a result, it is necessary to increase the discharge energy of the capacitive storage unit from transition to transition [6-9].

The remoteness of the impact, strict dosage of energy and the implementation of multiple loading are also needed in the implementation of the processes of calibration of metal elements (cuffs) of vacuum-tight MCU and straightening the bases of magnetic disks. The conditions of high-quality processing and technical requirements for special technological equipment (STE) are similar to those defined for the processing of composite MCU [9].

Comparison of theoretical studies of electromagnetic and power characteristics of magnetic pulse processing (MPP), data of high-speed photoregistration of pulse forming with data of measurement results and tests of the received details and knots allowed to define areas of effective variation of the parameters influencing quality of processing: working frequency of a discharge current f_d=3÷70 kHz; pulse energy W=10 kJ; working voltage U_w=0,5÷15 kV number of pulses in automatic mode N_t=1÷10. These data can be the source for the development of technical requirements for industrial equipment MPP [10-11].

4. Conclusion

To obtain composite integral axisymmetric compositions - metal, metal-glass and metal-ceramic units, magnetic pulse processing is expedient. It is possible to connect stranded wires with tips and splicing of onboard cable networks of aviation and space technology, assembly of metal-glass fuses and metal-ceramic units of electro-vacuum devices.

The use of uniform remote electromagnetic influence on the assembled units allows for more dense packing of conductive conductors in the contacts, eliminate soldering and provide heat dissipation in the insulation units of ultrahigh frequency devices.

References

[1] Sovetchenko P B 2003 *Possibilities of magnetic pulse processing of metals* (Tomsk: TPU publishing House)

[2] Badianov B N 2000 *Handbook of magnetic pulse treatment of metals* ed B N Badianov (Ulyanovsk: publishing house USTU)

IOP Conf. Series: Materials Science and Engineering 1005 (2020) 012016 doi:10.1088/1757-899X/1005/1/012016

[3] Glushchenkov V. A 2013 *Inductors for magnetic pulse processing of materials: tutorial* (Samara Educational literature) p 138
[4] Samokhvalov V N, Samokhvalova, Zh V and Grigoriev V L 2006 Experimental research and numerical modeling of the process of magnetic pulse *Assembly of multi-wire wires Bulletin of the Kurgan University. - Series Technical Sciences* **2.4.1** 104–06
[5] Strizhakov E L 2016 *Discharge-pulse processing of materials* ed E L Strizhakov, S V Nescoromniy, D V Minko (DSTU. Rostov-on-don : DSTU) p 201
[6] Yusupov R Yu 2013 *Power plants for magnetic pulse processing of materials* R Yu Yusupov, V A Glushchenkov (Samara: Publishing house "Fedorov") p 128 (in Russian)
[7] Strizhakov E L, Nescoromniy S V and Ageev S O 2016 *Welding International* **10** 813–816
[8] Strizhakov E L, Nescoromniy S V and Ageev S O 2015 *Welding International* **12** 988–990
[9] Kabluchev C V, Perlov D S, Rogozin D V and Strizhakov E L 2018 *Vestnik of DSTU* **1** 31–37
[10] Lemeshev S V, Strizhakov E L, Nescoromniy S V and Ageev S O 2015 *Symposium of technologists-machine builders* **1** 336–342 (in Russian)
[11] Strizhakov E L, Nescoromniy S V and Ageev S O 2016 *Abstracts of the report of the 5th International scientific seminar* **1** 57–59 (in Russian)

IOP Conf. Series: Materials Science and Engineering 1005 (2020) 012017 doi:10.1088/1757-899X/1005/1/012017

Structure and thermal diffusivity of SiC-NbC ceramics

Sh Sh Shabanov[1] **and G D Kardashova**[1,2]

[1] Department of Engineering Physics, Dagestan State University, Gadzhiev Street, 43-a, Makhachkala, the Republic of Dagestan, 367000, Russian Federation
[2] National Research Nuclear University MEPhI (Moscow Engineering Physics Institute), Kashirskoe shosse, 31, Moscow, 115409, Russian Federation

E-mail: gulya-ka11@yandex.ru

Abstract. The structure and thermal diffusivity of SiC - NbC ceramic materials are investigated . It is shown that with a change in the composition in sintered ceramics, the lattice parameter changes, and also in the near-contact region there is a mutual diffusion of silicon and niobium. This indicates the formation of solid solutions. As the niobium carbide content and temperature increase, the thermal diffusivity of SiC - NbC ceramic materials decreases.

1. Introduction

Ceramic based on silicon carbide has high mechanical strength, thermal conductivity, heat resistance and a low coefficient of thermal expansion. Due to the combination of these properties, SiC ceramics is already widely used in many branches of mechanical engineering, microelectronics and structures where there are high temperatures.

The addition of transition metal carbides to silicon carbide allows one to improve the mechanical properties and obtain ceramic materials based on silicon carbide with different electrical conductivities in the range from metal to semiconductor [1,2]. In particular, in the SiC – NbC system, the addition of niobium carbide to silicon carbide leads to an increase in fracture toughness up to 5.4 MPa·$m^{1/2}$ and Vickers hardness of 17.8 GPa.

2. Experimental Methods

Ceramic SiC - NbC was obtained by sintering at a temperature of 2373 K in an atmosphere of Ar for 1 hour. Powder SiC (green) polytype 6H was used. 5 μm SiC and NbC powders were mixed in a ball mill. The amount of SiC and NbC powder was: SiC (10%wt.)–NbC (90%wt.), SiC(30%wt.) – NbC (70%wt.), SiC (50%wt.) – NbC (50 %wt.), SiC (70%wt.) – NbC (30%wt.), SiC (90%wt.) – NbC (10%wt.). The density and porosity of the obtained SiC-NbC ceramic samples were determined by the method of filling and hydrostatic weighing. The density and porosity data for SiC - NbC ceramics are presented in Table 1. It can be seen that the density of the samples increases from $1.68 \cdot 10^3$ to $5.23 \cdot 10^3$ kg/m^3 with an increase in the NbC content from 10 to 90% by weight, and the porosity decreases.

The structure of the obtained ceramic materials SiC- NbC studied by the integrated intensity of the X-ray powder diffractometer model on Empyrean series 2 company PANalytical (X-ray tube voltage in the mode U=40kV, I=30mA), and scanning electron microscope «ASPEX» PSEM eXpress TM. Nickel filter, copper tube (Cu), 2 θ -geometry. The elemental

IOP Conf. Series: Materials Science and Engineering 1005 (2020) 012017 doi:10.1088/1757-899X/1005/1/012017

composition and distribution profile of the components in a thin layer near the boundary of the SiC - NbC systems was studied by Auger electron spectroscopy on an Auger - JAMP microanalyzer.

The thermal diffusivity was measured by a laser flash method (LFA 457 MicroFlach). Samples for measuring thermal diffusivity had dimensions: diameter – 12.7 mm, height – (2-3) mm.

Table 1. Density and porosity of SiC-NbC ceramics.

Temperature, K	NbC , %wt.	Density, 10^3 kg/m^3	Porosity, %
2373	10	1.68	53.7
2373	30	1.82	53.4
2373	50	2.29	52.4
2473	70	3.39	38
2373	90	5.23	17

3. Results and Discussion

Survey roentgenograms were taken in the range of Bragg angles $2\Theta = 15^0 \div 40^0$, with sample rotation, at maximum X-ray tube operating conditions and in the most sensitive range of reflection intensity measurements.

On the X-ray diffraction patterns, all lines corresponding to the X-ray spectra of both SiC and NbC are observed (comparison with tabular data on the ASTM file cabinet). No other lines corresponding to the formation of a new phase were found. To estimate the interplanar distances d (3,00) and d (00,18), a survey was performed in the range of angles $2\Theta = 59^0 \div 60^0$ and $2\Theta = 66^0 \div 68^0$. Interplanar distance calculated by the position of the line for wavelengths $\lambda_{\alpha 1}$ and $\lambda_{\alpha 2}$ separately. The error in the measurement d (3,00) and d (00,18) was ± 0.00004 Å and ± 0.00003 Å respectively. The results and measurements are shown in table 2.

In addition, in the X-ray diffraction patterns, individual weak reflexes are observed that are not repeated for samples of various compositions. When analyzing the angular positions of these lines, they cannot be identified as lines belonging to any compound of the type NbC, SiC_4, Nb_3Si_4, $NbSi_2$, Si_3Nb_5. It is seen from the X-ray diffraction patterns that, with a change in the NbC content, the intensity of lines corresponding to SiC decreases, and the intensity of lines corresponding to NbC increases.

Table 2. Results of x-ray diffraction studies.

NbC, %	1 shoot		II shooting	
	d(3,00)	d(00,18)	d(3,00)	d(00,18)
0	$d_{\alpha 1}$=0,8887	$d_{\alpha 1}$=0,8379	$d_{\alpha 1}$=0,8887	$d_{\alpha 1}$=0,8389
	$d_{\alpha 2}$=0,8883	$d_{\alpha 2}$=0,8378	$d_{\alpha 2}$=0,8884	$d_{\alpha 2}$=0,8392
10	$d_{\alpha 1}$=0,8881	$d_{\alpha 1}$=0,8386	$d_{\alpha 1}$=0,8881	$d_{\alpha 1}$=0,8389
	$d_{\alpha 2}$=0,8893	$d_{\alpha 2}$=0,8386	$d_{\alpha 2}$=0,8390	$d_{\alpha 2}$=0,8388
30	$d_{\alpha 1}$=0,8832	$d_{\alpha 1}$=0,8383	$d_{\alpha 1}$=0,8891	$d_{\alpha 1}$=0,8383
	$d_{\alpha 2}$=0,8895	$d_{\alpha 2}$=0,8387	$d_{\alpha 2}$=0,8890	$d_{\alpha 2}$=0,8387
50	$d_{\alpha 1}$=0,8897	$d_{\alpha 1}$=0,8385	$d_{\alpha 1}$=0,8853	$d_{\alpha 1}$=0,8385
	$d_{\alpha 2}$=0,8895	$d_{\alpha 2}$=0,8381	$d_{\alpha 2}$=0,8851	$d_{\alpha 2}$=0,8385
70	$d_{\alpha 1}$=0,8885	$d_{\alpha 1}$=0,8381	$d_{\alpha 1}$=0,8889	$d_{\alpha 1}$=0,8389
	$d_{\alpha 2}$=0,8892	$d_{\alpha 2}$=0,8380	$d_{\alpha 2}$=0,8890	$d_{\alpha 2}$=0,8384
90	$d_{\alpha 1}$=0,8887	$d_{\alpha 1}$=0,8384	$d_{\alpha 1}$=0,8880	$d_{\alpha 1}$=0,8388
	$d_{\alpha 2}$=0,8893	$d_{\alpha 2}$=0,8382	$d_{\alpha 2}$=0,8891	$d_{\alpha 2}$=0,8382

The identified lines from sample to sample are not reproduced. Since no additional lines indicating the formation of a new phase were found, it can be concluded from the analysis of X-ray diffraction patterns of SiC - NbC ceramics of various compositions that no phase transitions occur during sintering at a temperature of 2373 K and new compounds are not formed.

A study of the structural parameters “a” and “c” of the SiC unit cell was carried out by two-time shooting of each sample and when determining the parameters $k_{\alpha1}$ and $k_{\alpha2}$ separately, followed by averaging.

For SiC - NbC ceramics sintered at 2373 K, the parameter c decreases with an increase in the NbC content. The value of parameter “a” remains the same. A change in the parameter “c” implies a partial replacement of silicon atoms by Nb atoms in tetrahedrons.

In addition to the shift of the diffraction lines associated with a change in the unit cell parameters , an anomalous intensity distribution is observed in a series of reflections (00.18) from samples sintered at 2373 K, which manifests itself in ratios $J_{k\alpha1}$ and $J_{k\alpha2}$.

The line intensities $k_{\alpha1}$ and $k_{\alpha2}$ in the k_{α} doublet (theoretical and experimental) are referred to as 2:1. This ratio is associated with energy transitions that occur during the formation of characteristic x-ray radiation.

The ratio $J_{k\alpha1}/J_{k\alpha2}$for ceramic samples with 10% weight. NbC is 2:1. As SiC is replaced by NbC , the line intensity $J_{k\alpha1}$ gradually decreases, while it $J_{k\alpha2}$ increases.

It should be noted that the study of the distribution of the intensity profile reveals a very diverse distortion of the crystal lattice. Moreover, depending on the type of violation in the lattice, diffraction effects can manifest themselves on reflections of different types in different ways. In this case, such a change in intensity is presumably associated with one-dimensional disordering of SiC. Since the SiC structure has a polytypism, it can be assumed that the disordering of densely packed SiC layers occurs. In this case, a violation of the order inside the layer does not occur.

With the change in the percentage of SiC and NbC to ceramics sintered at 2 to 373 K, there is a one-dimensional disordering packed layers SiC along the axis "c".

In Figures 1-4 shows the experimental data on the temperature dependence of the thermal diffusivity of SiC - NbC ceramics of various compositions (10-90% wt.) NbC.

With increasing temperature and niobium content, thermal diffusivity decreases. This may be due to the microstructure of SiC - NbC ceramic materials.

In Figure 5 shows the microstructure of a ceramic sample with SiC composition (70%) - NbC (30%) and composition spectra at points 1,2 and 3, respectively. Point 1 corresponds to a grain of silicon carbide, point 2 corresponds to niobium carbide, and point 3 is an intermediate phase, in which silicon and niobium are located. This indicates that solid-phase diffusion occurs in the contact zone.

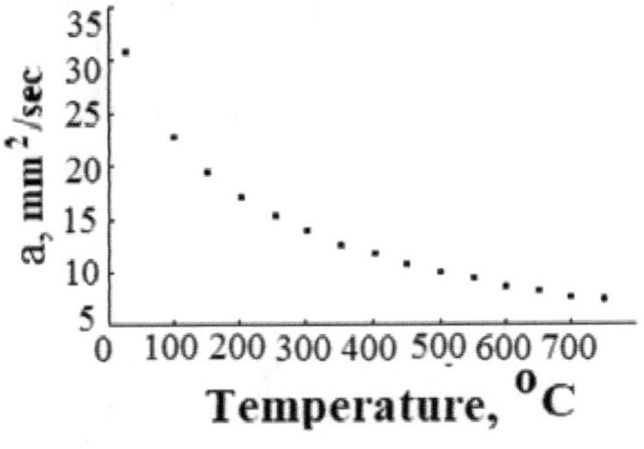

Figure 1. The dependence of thermal diffusivity of ceramics SiC(90%)-NbC(10%).

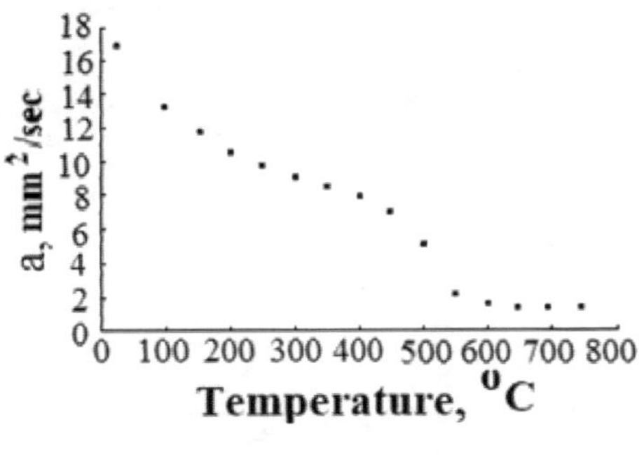

Figure 2. The dependence of thermal diffusivity of ceramics SiC(70%)-NbC(30%).

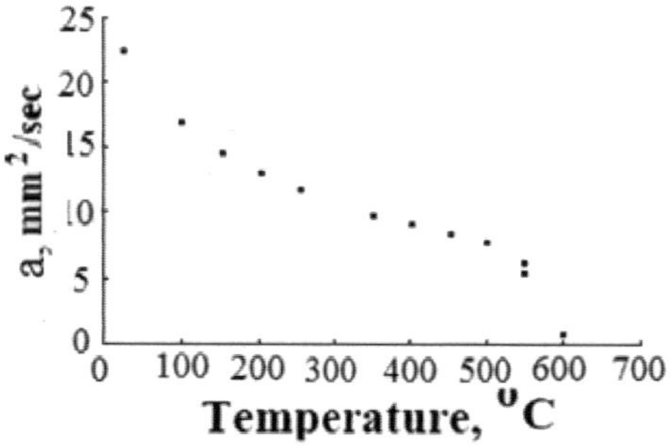

Figure 3. The dependence of thermal diffusivity of ceramics SiC(50%)-NbC(50%).

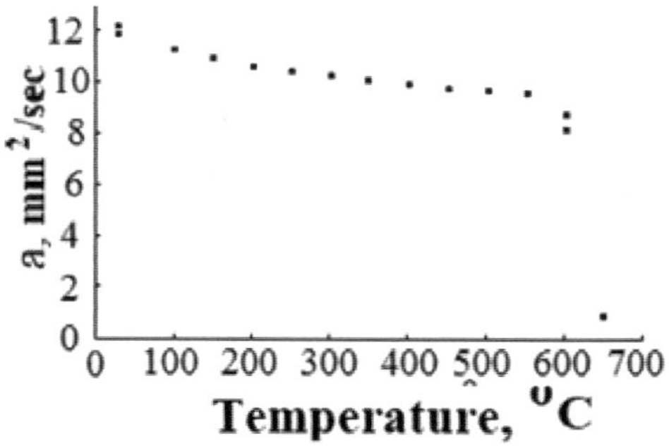

Figure 4. The dependence of thermal diffusivity of ceramics SiC(10%)-NbC(90%).

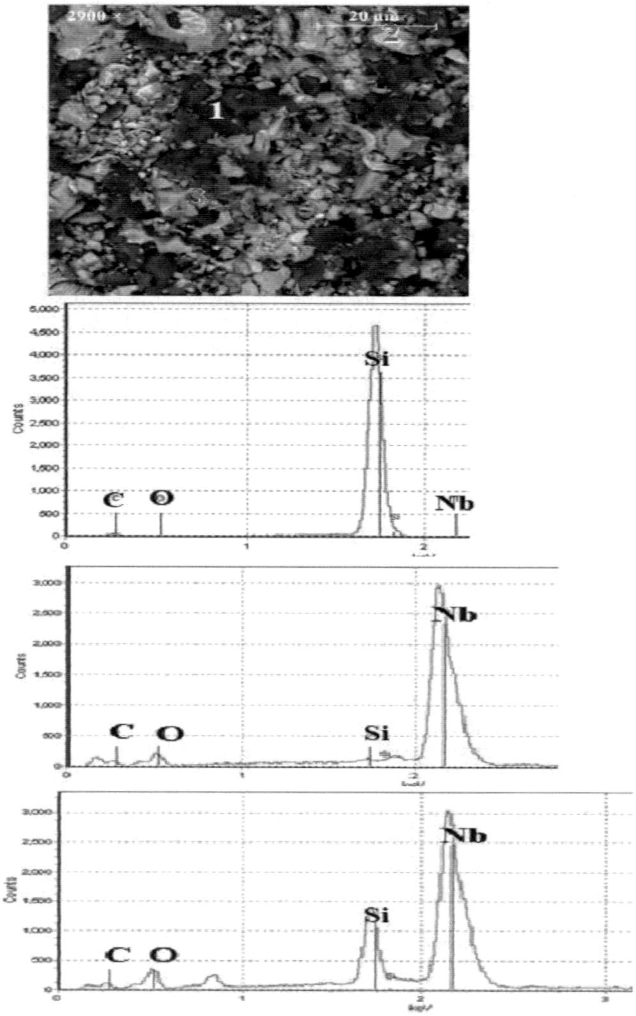

Figure 5. Microstructure of SiC - NbC ceramics (30%) and spectra at points 1, 2, and 3, respectively.

In Figure 6 shows the Auger spectra of a single crystal of NbC (a), a single crystal of SiC (b). Along with the peaks of silicon and carbon, a Nb peak is observed in the SiC single crystal (Figure 6 (c)), which confirms the presence of active mutual diffusion in the SiC - NbC system during solid phase sintering.

In Figure 7 shows Auger profiles of atomic concentrations of C_{Nb} in SiC and C_{Si} in NbC at T = 2420 K and P = 40 MPa.

According to the concentration profiles of Nb, in SiC and Si in NbC, it can be seen that they are similar and concentrated near the interface, and a continuous increase in the partial atomic concentrations of C_{Nb} leads to a decrease in C_{Si} and vice versa (Figure 7). The observed behavior of the partial atomic concentrations of C_{Nb} in the SiC and C_{Si} in NbC, indicating the formation of a continuous series of solid solutions in the contact area of SiC to NbC. It is seen that the solubility of Nb in SiC is higher than the solubility of Si in an NbC single crystal. A detailed analysis of the interface showed the absence of second-phase inclusions and chemical compounds that occur in the SiC-NbC system.

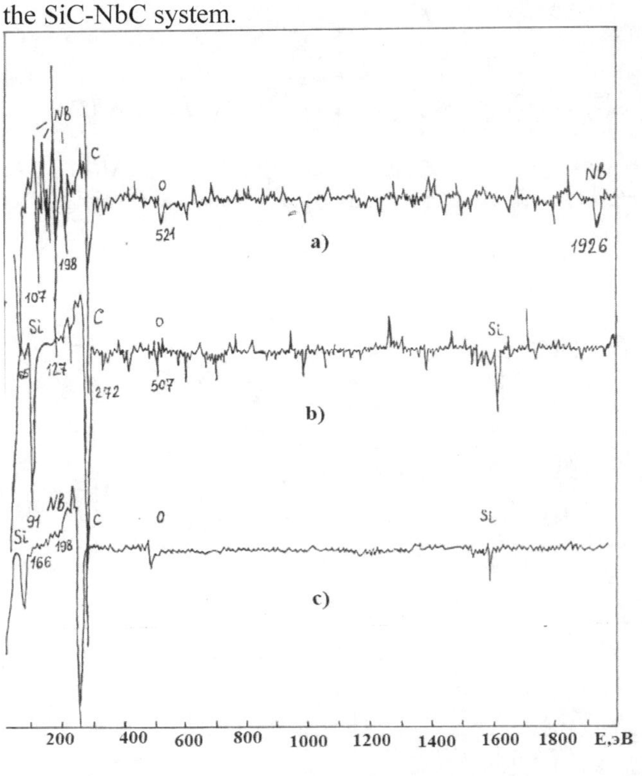

Figure 6. Auger spectra of single crystals of NbC (a), SiC (b) and in solid SiC - NbC solution at a point spaced 1 μm from the boundary (c)

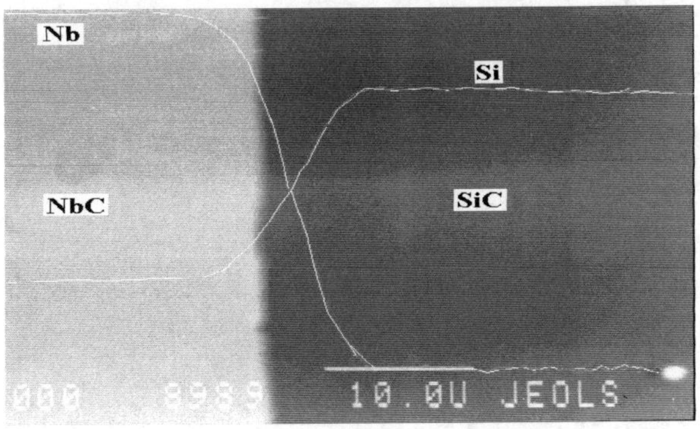

Figure 7. Concentration profiles of Nb in SiC and Si in NbC at T = 2373 K (Auger analysis)

4. Conclusion

X-ray diffraction analysis of SiC - NbC ceramic materials of various compositions showed that phase transitions do not occur during sintering at a temperature of 2373 K and new compounds are not formed. For SiC - NbC ceramics sintered at 2373 K, the parameter c decreases with an increase in the NbC content , which indicates the formation of solid solutions. With the toast of temperature and NbC content in SiC - NbC ceramic materials sintered at a temperature of 2373 K, the thermal diffusivity decreases. In the near-contact region, solid-phase diffusion occurs between the SiC and NbC grains.

Acknowledgements

This work was supported by the Ministry of Education and Science of the Russian Federation, project № 11.1957.2017/4.6.

References

[1] Safaraliev G K, Shabanov Sh Sh, Bilalov B A and Sadykov S A 2011 *Herald of the DSU* **6** 31–34 (in Russian)

[2] Safaraliev G K, Kardashova G D, Bilalov B A, Shabanov Sh Sh, Vagabova G A and Aliev M A 2013 *Fundamental problems of electronic instrumentation* **13** (2) 126–129 (in Russian)

IOP Conf. Series: Materials Science and Engineering 1005 (2020) 012018 doi:10.1088/1757-899X/1005/1/012018

Mechanical properties of ceramic materials SiC-BeO

Sh Sh Shabanov[1] and G D Kardashova[1,2]

[1] Department of Engineering Physics, Dagestan State University, Gadzhiev Street, 43-a, Makhachkala, the Republic of Dagestan, 367000, Russian Federation
[2] National Research Nuclear University MEPhI (Moscow Engineering Physics Institute), Kashirskoe shosse, 31, Moscow, 115409, Russian Federation

E-mail: gulya-ka11@yandex.ru

Abstract. High-density ceramic based on silicon carbide with the addition of beryllium oxide compositions (1-2)% wt. With a step of 0.2 was obtained. X-ray diffraction studies of SiC-BeO ceramics showed changes in the lattice parameter of SiC "c" with the addition of beryllium oxide, the minimum value of which was observed for a composition of 1.4% wt. BeO. For compositions greater than 1.4% wt. BeO was observed only polytype 6H. The grain size of sintered ceramic materials reached up to 20 microns. The elastic moduli, microhardness, and shear strength for hot-pressed SiC-BeO ceramic materials are determined.

1. Introduction

Ceramics based on silicon carbide, due to its useful properties, is one of the most promising materials for various branches of mechanical engineering. It has high mechanical strength, resistance to oxidation at high temperature, high heat resistance and thermal conductivity. However, the implementation of these useful properties is possible upon receipt of high-density silicon carbide materials. One of the modern technological methods for producing high-density ceramic materials based on silicon carbide is hot pressing [1-4].

Due to the covalent nature of the Si-C bond, sintering of silicon carbide without the use of sintering activators ends when a density of about 70% of theoretical (TP) is reached, which sharply reduces the strength properties of ceramics.

It is known that materials with high solubility in SiC and forming solid solutions with it are an effective additive for sintering silicon carbide. Among these compounds are, in particular, beryllium oxide having a wurtzite structure characteristic of alpha-modification of silicon carbide and, as a result, forming solid solutions [5].

Ceramic materials SiC-BeO are interesting in a combination of high thermal conductivity and electrical resistance [6]. Adding a small amount of beryllium oxide (~ 2% wt.) To silicon carbide leads to an increase in thermal conductivity up to 270 W m·K, and a specific resistance of 10^{13} Ω·cm.

2. Materials and Experimental Methods

SiC-BeO ceramics was obtained from green powder of 6N modification silicon carbide, with a dispersion of 2.4 microns with the addition of beryllium oxide powder with a fineness of 1 micron, by hot pressing at a temperature of 2420 K and a pressure of up to 35 MPa for 1 hour in an atmosphere of N_2. SiC and BeO powders were mixed in a ball mill for 8-24 hours. The shear strength was measured by the four-point method, and the microhardness was indented by the Vickers prism. To measure the flexural strength, the samples were prepared in the form of beams (7x7x70) mm. Microhardness was studied on samples of silicon carbide ceramics with BeO additives in the form of a cylinder with a diameter of 20–40 mm and a thickness of 10 mm. The surface of polycrystalline samples was previously ground and polished with a diamond tool.

We calculated the elastic moduli from the measured values of the density and propagation velocities of the longitudinal C_l and transverse C_t ultrasonic waves. The propagation velocity of ultrasound is measured by pulsed and resonant methods. Pulse measurements were carried out using an acoustic tester in the "On passage" mode at a frequency of 5 MHz at a temperature of 20 ± 2 °C. The frequencies of natural bending and radial vibrations of the disks were measured according to the method described in [7]. The speeds C_l and C_t are calculated from these frequencies. We used samples in the form of washers with a diameter of 20 mm and a thickness of 7 mm.

The structure of the obtained ceramics was studied by the integrated X-ray intensity of CuK_α on a DRON-2.0 diffractometer at an accelerating voltage of 20 kV. X-ray structural analysis of SiC-BeO-based ceramic samples was carried out both on pressed washers and on powders made from hot-pressed samples. Powders were made by abrasion of pieces of samples in a sapphire mortar. The results of the decoding of the X-ray diffraction patterns were compared with the two closest polytypes of SiC (data taken from the ASTM file cabinet). The microstructure of SiC-BeO ceramics was studied by cleavage (without selective etching) using a JSM-T300 microscope.

3. Results and Discussion

Density data for the hot-pressed silicon carbide ceramic SiC-BeO are presented in table 1.

A general regularity for SiC-BeO ceramics obtained from -SiC powder is to increase the matrix density to certain composition values (~ 2% by weight). This behavior of the density dependence on the composition for SiC-BeO ceramics is associated with two simultaneously proceeding processes: filling the intergranular phase with beryllium oxide and dissolving beryllium oxide in silicon carbide during the diffusion interaction of SiC and BeO powders.

Table 1. Density and polytype composition of SiC-BeO ceramics

№№	Density ρ, g/cm^3	Porosity P, %	Polytype	Temperature T, K	Composition, weight % BeO
1.	2,77	13,7	6H,15R	2420	1
2.	2,78	13,4	6H,15R	2420	1,2
3.	2,815	12,3	6H,15R	2420	1,4
4.	2,795	12,53	6H	2420	1,5
5.	2,85	11,2	6H	2420	1,6
6.	2,85	11,2	6H	2420	1,8
7.	2,98	7,16	6H	2420	2

Figure 1 presents a spectrometric analysis of the dispersion of energy of a ceramic SiC-BeO composition (2% wt. BeO). The analysis showed the presence of a phase of silicon, carbon, oxygen and the absence of beryllium.

The study of the parameters of the crystal structure of SiC-BeO on the composition showed that in the range of (0.5-2) % wt. The peaks in the diffraction patterns inherent to pure SiC were shifted by $(0.1–0.2)^\circ$ toward small angles, which indicates a change in the crystal lattice parameter. This indicates the formation of a solid solution of SiC-BeO. From the dependence of the unit cell parameter "c" of SiC-BeO ceramics on the composition, it was found that the minimum value of the parameter is observed for samples with additives of (1.4-1.5) % wt BeO. As shown by the results of x-ray phase analysis, under conditions of activated sintering in the presence of beryllium oxide, polytype transitions 6H, 15R→6H occur.

The microstructure of SiC-BeO ceramics was studied by cleavage (without selective etching) using a JSM-T300 microscope (Figures 2,3).

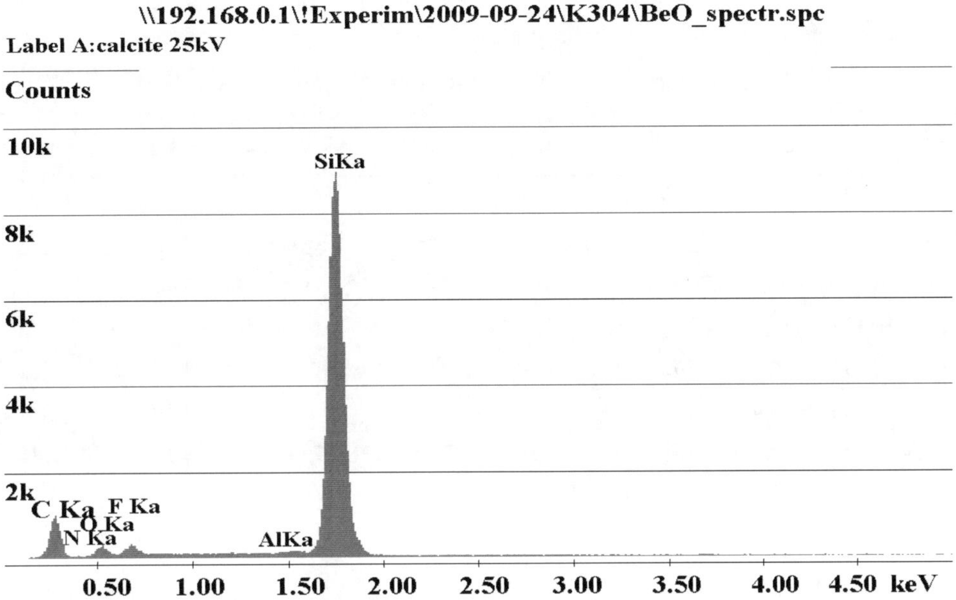

Figure 1. Spectrometric analysis of the energy dispersion of SiC-BeO ceramics (2% wt. BeO).

The structure of sintered SiC-BeO represented α-SiC grains up to 20 μm in size. The study of the microstructure showed that sintering is activation in nature, due to the mechanical activation of the powders. The content of beryllium oxide ~ (1.4–1.6% wt.) is optimal for obtaining the most dense material with a uniform fine-grained structure.

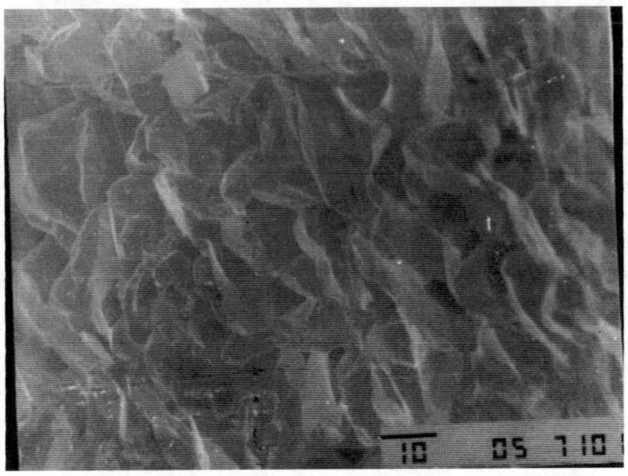

Figure 2. Microstructure of SiC-BeO Ceramics (1.4 wt% BeO)

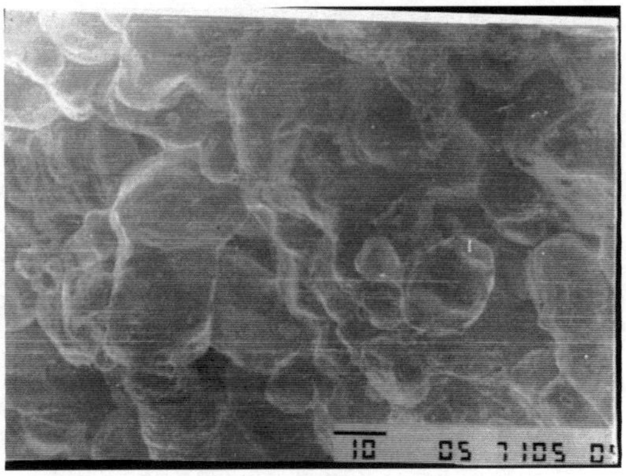

Figure 3. Microstructure of SiC-BeO Ceramics (2,0 wt% BeO)

The results of measurements of the longitudinal and transverse velocities of ultrasound and elastic moduli for SiC-BeO ceramic materials are presented in Table 2 and Figure 4.

Figure 4 shows the correlation field of Young's modulus (E) and porosity obtained on several batches of samples of silicon carbide ceramics. The confidence interval with a probability of 0.95 contains unaccounted systematic and random errors in measuring the propagation time of ultrasound in the sample and the length of the sample. A study of the elastic moduli of SiC-BeO ceramic materials showed that porosity is the main factor affecting their elastic properties. In [1,4], a Young's modulus of 513 GPa was obtained in samples of theoretical density at room temperature. In our samples, the porosity of which varies from 7% to 13%, small additives of beryllium oxide (up to 2% wt. BeO) do not strongly affect the elastic properties.

Table 2. Elastic Modules of SiC-BeO Ceramic Materials

№№	Cl, m/c	Ct , m/c	μ	E, GPa	M, GPa	G, GPa	wt % BeO
1	11527.6+110	7373+30	0.153	374	180	162	2
2	11252.0+80	7229+30	0.148	345	163.35	150	1.8
3	10909.0+90	7034+20	0.145	322	151.17	141	1.6
4	10532.95+80	7272+25	0.131	333	150.406	147	1.5
5	11293.0+110	7293+50	0.143	342	153.664	150	1.4
6	10874.0+110	7071+30	0.136	316	144.7	139	1.2
7	10898.7+150	7058+50	0.136	316	145.03	138	1
8	9196.6+90	6030+80	0.124	253	91.3	92	0

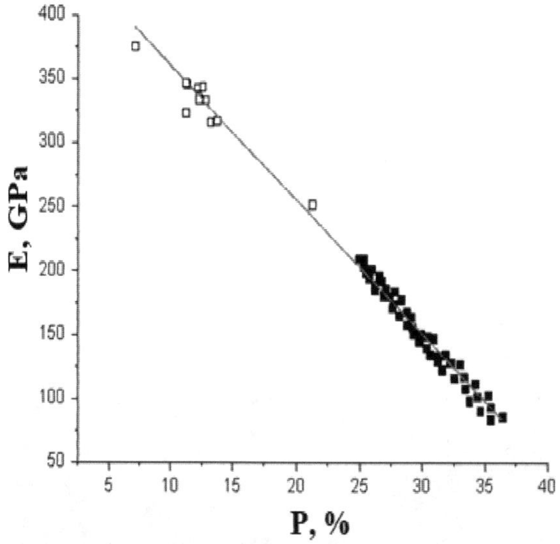

Figure 4. The dependence of the Young's modulus in SiC-BeO ceramics on porosity: □ - hot-pressed ceramic materials SiC-BeO; ■- batch of samples of nitrided silicon carbide (46 samples in total).

Figure 4 shows that the elastic modulus for SiC-BeO samples decreases in SiC-BeO ceramics with increasing porosity. A good correlation of the elastic moduli with porosity allows the use of acoustic measurements to expressly evaluate the porosity of the resulting SiC-based ceramic. On the other hand, knowing the elastic properties of the sample and its porosity, it is possible to evaluate the elastic properties of the ceramic matrix and draw conclusions about the influence of a particular impurity on its properties.

In the study of microhardness by the indentation method, no cracks were observed on the fingerprints, due to the presence of porosity. The maximum microhardness falls on the composition region (1.4–1.5) % wt. BeO and amounted to 17 GPa. This indicates a change in the bond strength between the atoms, which is usually observed during the formation of a solid solution. This is also confirmed by a decrease in the lattice parameter in SiC-BeO ceramics, and by attaining a minimum in the region of SiC concentration + 1.4 wt. %. BeO.

The shear strength data are presented in Figure 5. It can be seen that, up to a temperature of 1300 K, the tensile strength remains constant.

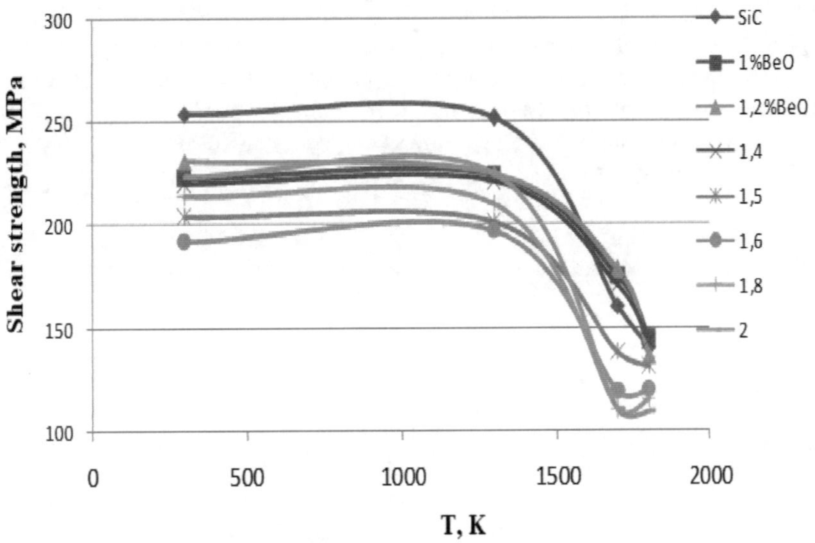

Figure 5. The temperature dependence of the tensile strength of hot-pressed ceramic SiC-BeO of various compositions on temperature.

The shear strength at temperatures of 300 K and 1700 K for ceramic compositions SiC + 1.4% wt. BeO and SiC compositions + 1.5% wt. BeO was 230 MPa and 160 MPa, respectively. The fall in the value of the shear strength of ceramics with the composition SiC + (1.4–1.5)% wt. BeO less than pure hot pressed SiC Moreover, the value of the ultimate shear strength in ceramics of SiC + compositions is (1.4–1.5)% weight. BeO at a temperature of T = 1770 K is retained and even a slight increase is observed in comparison with pure silicon carbide. The effect of the addition of beryllium oxide on the value of the shear strength is not as great as in the case of microhardness.

4. Conclusion
Dense ceramic was obtained in the SiC-BeO system by hot pressing. X-ray diffraction studies of SiC-BeO ceramics showed changes in the lattice parameter on the composition of beryllium oxide, as well as the activation character of sintering of SiC-BeO ceramics. The grain size as a result of recrystallization increased to 20 μm. The main factor influencing the elastic properties of SiC-BeO ceramics is porosity. The maximum microhardness was observed at concentrations of 1.4-1.5% wt. BeO and amounted to 17 GPa. The shear strength remains constant up to a temperature of 1300 K and varies between 200 and 250 MPa.

Acknowledgements
This work was supported by the Ministry of Education and Science of the Russian Federation, project № 11.1957.2017/4.6.

IOP Conf. Series: Materials Science and Engineering 1005 (2020) 012018 doi:10.1088/1757-899X/1005/1/012018

References

[1] Neuman Eric W, Hilmas Gregory E and Fahrenholtz William G 2015 *J. European Ceramic Society* **35**(2) 463–476
[2] Šajgalík P, Sedláček J, Lenčéš Z, Dusza J and Lin H-T 2016 *J. European Ceramic Society* **36**(6) 1333–41
[3] Lim Kwang-Young, Kim Young-Wook and Kim Kwang Joo 2014 *Ceramics International* **B 40**(7) 10577–82
[4] Neuman Eric W, Hilmas Gregory E, and Fahrenholtz William G 2013 *J. European Ceramic Society* **33**(15-16) 2889–99
[5] Safaraliev G K, Shabanov S S, Kardashova G D , Akhmedov R R, Al-Heyun F G and Emirov S N 2016 *Bulletin of the Russian Academy of Sciences: Physics* **80**(5) 500–503
[6] Safaraliev G K, Shabanov S Sh, Bilalov B A, Kardashova G D, Abilova N A and Abdeljavad X A 2008 *News of higher educ. Instit. Mater. of electronic equipment* **2** 34–36 (in Russian)
[7] Miloserdin Yu V and Baranov V M 1978 *M.: Atomizdat* (in Russian)

IOP Conf. Series: Materials Science and Engineering 1005 (2020) 012019 doi:10.1088/1757-899X/1005/1/012019

Pulsed powder technologies for the production of permanent compounds from heterogeneous materials

V Vinogradov, E Strizhakov and S Nescoromniy

Don State Technical University
Gagarin square 1, Rostov-on-don, 344000, Russia

E-mail: Vbif001@yandex.ru

Abstract. A study of the process of high-speed consolidation of powder materials with simultaneous formation of a welded joint with the surface of the metal base is presented. Thermal and power effects are used as a result of passing a current pulse through the powder composition and the inductor. The electric scheme of serial connection of a zone of processing and the magnetic hammer consisting of the flat inductor and the pusher is offered. The device of electric pulse welding-pressing is connected to the pulse current generator with a capacitive energy storage. Pulsed currents with amplitude of 200 – 300 kA allow to obtain Joule heat, warming up the composition to premelting temperatures (0.6 – 0.8) $T_{p.}$, and magnetic pressure of $500 \cdot 10^6$ N / mm^2 during the process (100 – 200) $\cdot$ 10^{-6} s. energy-intensive equipment up to 10 kJ with a discharge current frequency of 5 kHz is used. The study of the connection zone showed that the welding of the composition with monolithic materials occurs in the solid phase, no new structures are formed. The introduction of a monolithic material in the form of a wire into the pressing zone allows to obtain a connection of different materials with the preservation of the original structure. The powder composition is in this case the connecting element. Thus, copper compounds with steel, aluminum, brass and other materials are obtained. The dependences of the strength of compounds on the discharge energy of pulse current generator are obtained.

1. Introduction

Despite significant difficulties (crack formation, due to the difference in the coefficients of linear expansion of metals, intermetallic compounds) consolidation of heterogeneous materials, their use is growing in modern technology. This is due to the significant technical and economic advantages that have designs of dissimilar materials. The joints are realized in different design solutions and in different welding methods. In recent years, special attention has been paid to methods of joining dissimilar materials using powder compositions [1].

Consolidation of powders by traditional methods is due to the complexity of the process and takes considerable time. The authors of the article set a goal – to develop and investigate the connection of monolithic heterogeneous materials using pressed powder compositions as a binder. Consolidation is proposed to be carried out by pulsed electric heating, which significantly reduces the time of obtaining the connection.

2. Electro-welding-pressing with a static load

As the object of the study, the structural scheme the connection copper wire with a steel base-an analogue the design of the output electrochemical protection (ECP) pipelines Figure 1 [2].

Traditionally, the output of ECP is obtained by thermite welding. In the process of combustion of the thermite mixture, a copper-based melt is formed, which is the link in the connection of steel and copper. The authors propose to replace the melt with compressed and sintered powder. It is proposed to carry out this by pulsed electrical equipment [3,4].

Electric pulse welding-pressing (EPWP) is a method of obtaining compact products from powders, which use the simultaneous impact on the powder billet of a powerful high-voltage pulse of electric current and mechanical pressure.

The discharge of the pulse current generator (PCG) provides a powerful energy release in powder pressing. [5,6].

In the EPWP process, the sintered powder billet is compressed by the forces of the magnetic field created by the electric current and the pressure from the external loading system [7].

The short pulse duration at an EPWP of $(100 - 200) \cdot 10^{-6}$ s provides a high rate of compaction of the powder material, which makes it possible to carry out the process in air without the use of a protective atmosphere or vacuum. High-speed consolidation if EPWP enables to keep the pressing properties and the composition of the starting powder. Pulse heating with a static pressure of 10 kg / mm^2 allows to obtain products with a material porosity 20 – 30% [8].

For the manufacture of the structure presented in Figure 1 the authors proposed a device for obtaining compounds from heterogeneous materials, shown in Figure 2 [9]. At EPWP the matrix in diameter of 10 mm, with height of filling of a powder of 5 mm, diameter of the electrode-punch of 10 mm was used. Quality was estimated by mechanical tests on shift, it reached values F_s=170 N, Figure 6.

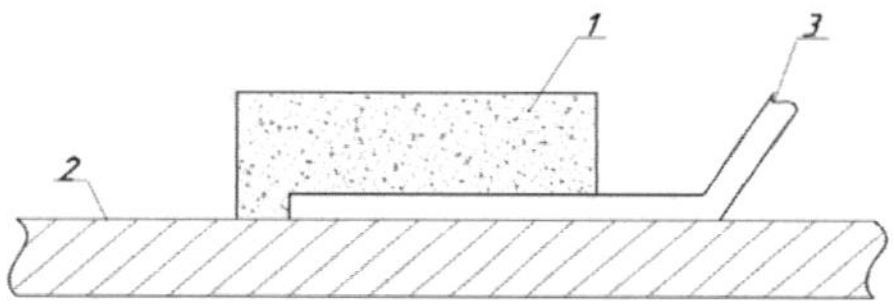

Figure 1. The constructive scheme of a conclusion of ECP. 1 – powder pressing (instead of sintered thermite mixture); 2 – metal plate (pipe simulator); 3 – lead (copper wire).

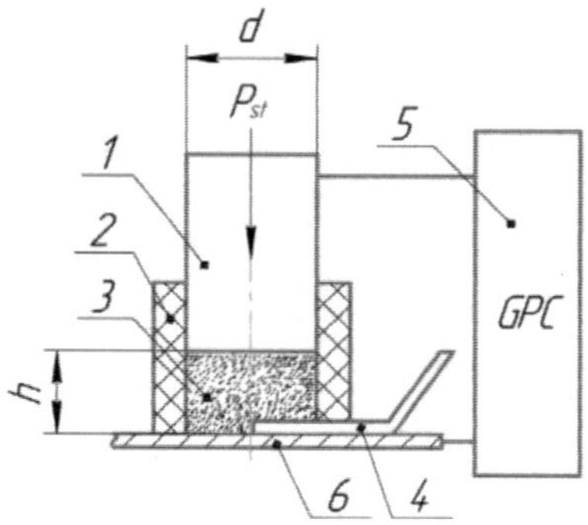

Figure 2. Device for obtaining compounds of dissimilar conductive materials by EPWP 1 – electrode punch; 2 – dielectric matrix; 3 – powder; 4 – wire; 5 – PCG; 6 – metal plate; P_{st} – static load; d-electrode diameter; h – height of powder filling.

3. Electrical impulse consolidation of materials with induction-dynamic loading

The authors of the article proposed the use of electric pulse consolidation of powder materials with induction-dynamic loading [10].

The essence of this method is to act on the powder with a short high-voltage pulse of current and mechanical pressure in the form of electromagnetic dynamic loading. The principal difference of this method is the use of induction-dynamic drive (IDD), included in the circuit with a punch-electrode and powder, as a result of which it is possible to simultaneously produce pressing and sintering of the material, using one generator of pulse currents with synchronization of mechanical and thermal effects on the powder [11].

The scheme of the device is shown in Figure 3. The powder material 1 is placed in a matrix 2 of non-conductive material. Electrode-punch 3 transmit pressure to the powder blank 1 and simultaneously serve as current leads from the PCG to the pressed roller [12].

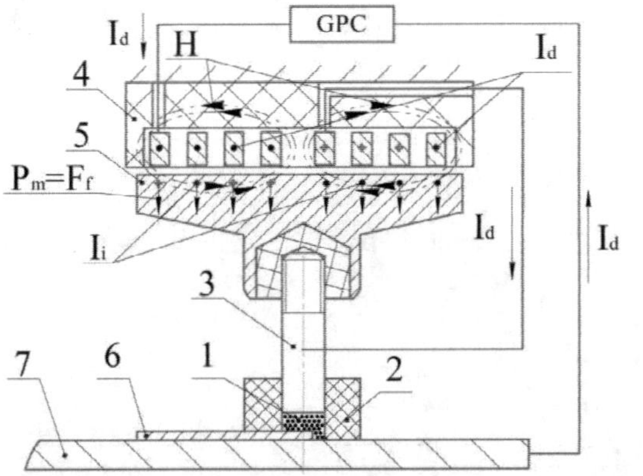

Figure 3. Schematic diagram of the EPWP process with IDD: 1 – powder; 2 – dielectric matrix; 3 – punch-electrode; 4 – inductor; 5 – pusher; 6 – wire; 7 – plate; I_d – discharge current; I_i – induced current; H – magnetic flux; P_m – magnetic pressure; F_f – forging force.

When discharge PCG high-frequency discharge current I_d flows through the inductor 4, the electrode punch 3 and the plate 7. Thus, the thermal effect on the powder 1 is carried out. In the working area of the inductor 4, an alternating magnetic field H is formed, it induces induced currents I_I in the pusher 5. The interaction of I_I with the magnetic flux H leads to the emergence of electromagnetic forces-magnetic pressure P_m., which is converted into mechanical forging force F_f, which compacts the powder composition and contributes to the pressing and welding process.

A current pulse with a density of about 10^4 A / cm^2 passing through the punch electrode and the powder intensely heats only the powder material without significantly heating the punch. This is due to the fact that the resistivity in the contact zone of the particles of the powder material significantly exceeds the resistivity in the contact zone of the punch-powder. The height of powder filling mainly depends on granulation and chemical composition and remains constant for specific types of powders [13 – 15]. Pulse heating with induction-dynamic drive makes it possible to obtain products with a material porosity of 5 – 10% [15 – 16].

The synchronization of mechanical and thermal effects is illustrated in Figure 4 the dependence of the discharge current and magnetic pressure of the IDD on time against the background of voltage changes on the capacitive accumulator PCG and the movement of the electrode-punch 3.

The amplitude of the magnetic pressure was about $500 \cdot 10^6$ N / mm^2, the current frequency is 5 kHz, the process time is $180 \cdot 10^{-6}$ s. [15, 17, 18].

The discharge voltage and current are damped harmonic oscillations. The coincidence of the current heating period with the action of magnetic pressure is a consequence of the serial connection of the IDD and the electrode to the energy source [15, 19 – 22].

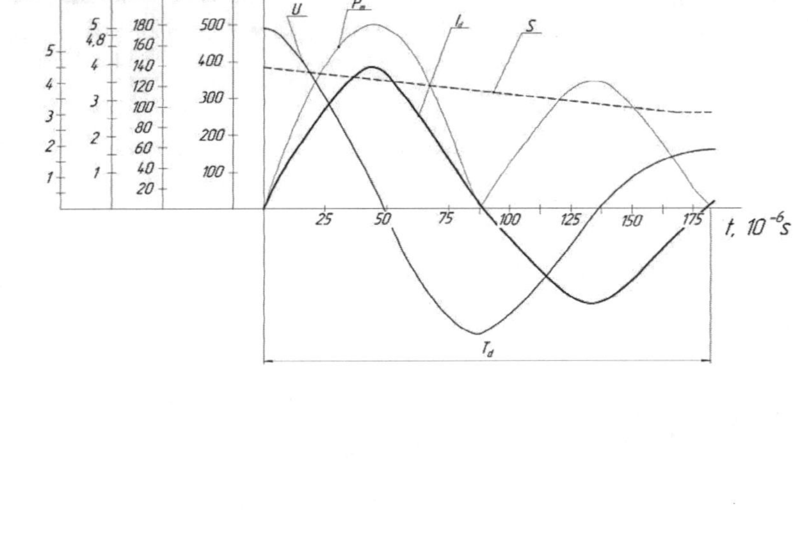

Figure 4. The relationship of the parameters of the process of EPWP with IDD U – voltage; P_m – magnetic pressure; I_d – discharge current; T_d – the period of discharge current; S – movement of the electrode-punch.

Heat and force is due to the energy stored in PCG with the capacitor capacitance C. Without losses in current supply, the amount of heat and power impact and the total stored energy W in the device is described by equation (1):

$$W = \frac{CU^2}{2} = UI_dT_d + \frac{\mu H^2}{2} \qquad (1)$$

where: C – storage capacity; U – the charge voltage of the capacitor; I_d – discharge current; T_d – discharge current period; μ – magnetic constant; H – magnetic field strength.

For Figure 5 presents an experimental sample welded EPWP with IDD, which was tested by a shear force from the value of the input energy.

The shear force dependencies are shown in Figure 6. Compounds produced EPWP with IDD have a greater mechanical strength than EPWP with static pressure by 15 – 20%. In the energy range of 2.8 – 4.8 kJ, compounds of unsatisfactory quality are formed, this is due to the short duration of the thermal effect. And in the energy range of 5.5 – 7.2 kJ, splashes of molten powder particles appear, which worsen the quality of the compounds. In the energy range of 4.8 – 5.5 kJ, a compound with a maximum strength of F_s =200N is formed between the sintered powder and the surface of the monolithic metal.

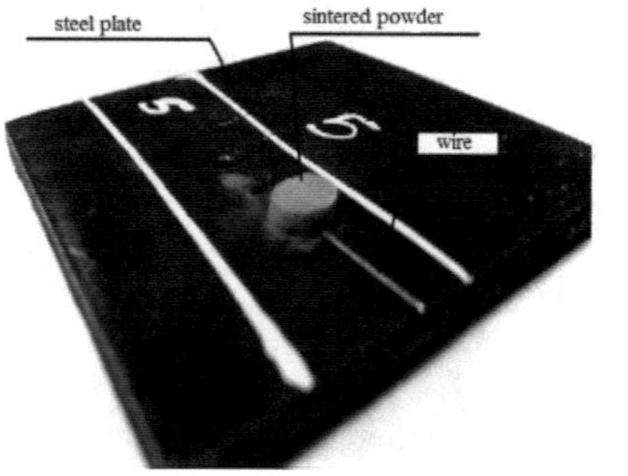

Figure 5. The experimental sample of a compound copper wire with a steel plate by using copper powder.

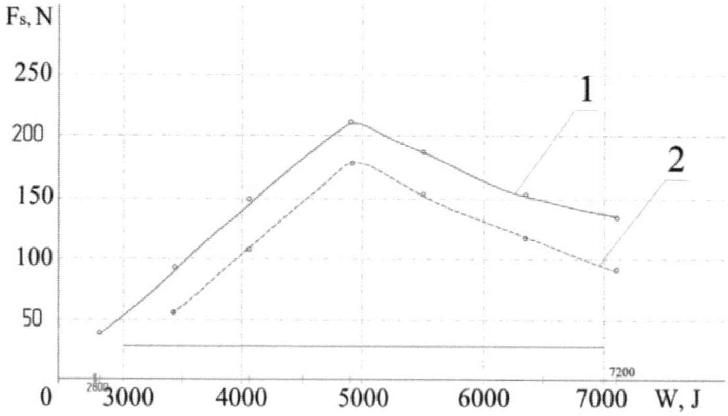

Figure 6. Dependence of the shear force F_s on the value of the energy storage W: 1 – with the use of IDD; 2 – with static pressure.

Compounds of dissimilar materials were obtained by EPWP with IDD. In table. 1 modes of obtaining qualitative connections of various materials with copper wire of the C1k brand with a diameter of 1,2 mm with use of various powders are presented. The same parameters used when EPWP with static loading (diameter of the matrix of the electrode plug and the height of the backfill of powder).

Table 1. Energy (kJ) required to obtain a high-quality connection of various materials.

Powder	The plate material				
	Copper CN 95-5	Aluminum AD1	Steel BSt3cm	Brass B95	Steel R18
Copper PMC – K	3	3,5	5	4	4,5
Aluminum PA – 1	3	3	3,5	3,75	4,5
Nickel PNK – UT4	3,5	3,5	3,5	4	4,5

For Figure 7 presents the micro-sections and microslip compounds of different materials with copper wire C1k diameter of 1.2 mm.

Studies of micro-grinding show that a clear interface is visible between the powder pressing and the monolithic metal, in which no intermetallides and other inclusions are detected, which indicates the formation of the compound in the solid phase. On macrolife can be seen that the deformed wire has an oval shape, is the result of the efforts of the forging F_f to induction-dynamic drive.

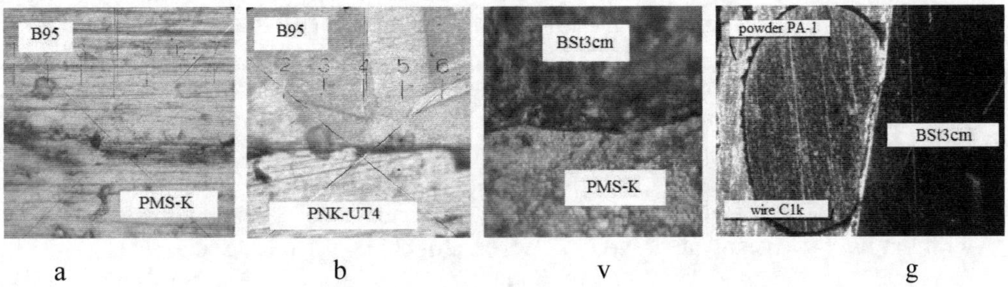

Figure 7. Micro-characters (a, b, v) ×485, macro-characters ×20 (g):
a – connection of copper powder brand PMC – K with brass plate brand B95; b – connection of nickel powder brand PNK – UT4 and brass plate brand B95; v – connection of copper powder brand PMS – K and steel plate brand BSt3cm; g – connection of aluminum powder brand PA – 1 and steel plate BSt3cm with copper wire C1k.

The technology of obtaining a compound of heterogeneous materials using pulsed current, provided high quality sintering of the powder with the simultaneous formation of welding between the powder and the surface of monolithic metals.

4. Conclusion

The connection of dissimilar monolithic materials can be carried out using as a binder a powder composition obtained by electric pulse pressing. This allows, in comparison with other processes, to increase the productivity of consolidation, ensuring the preservation of a pre - organized structure of the materials to be joined. The devices providing heating of a zone of connection by pulse currents with static and shock mechanical influence are created. Synchronization of thermal and induction-dynamic action is carried out by serial connection of heating elements and magnetic hammer to the generator of pulse currents.

It is established that the quality of compounds obtained by electric pulse pressing - welding with pulse load is higher than when using static action.

Compounds of ferrous and non-ferrous materials were obtained by the EPWP method in various combinations. Conclusions on the work, recommendations on the use of the results are presented.

References

[1] Lysak V I 2015 *Compaction of powders by explosion* (Moscow: Mashinostroenie) p 252 (in Russian)

[2] STO Gazprom 2-2. 2-136-2007 2007 *instruction on welding technologies in the construction and repair of field and main gas pipelines, part 1* No. 22-09-2007 (M.: Publishing house "Scientific research Institute of natural gases and gas technologies-VNIIGAZ) (in Russian)

[3] Baranov Y V 2001 *Physical principles of electropulse and electroplastic treatments and new materials* (Moscow Min. Blagonravova, MGIU) p 843 (in Russian)

[4] Belyavin K E 2004 *Modeling of the process of electric pulse sintering of metal powders* (Engineering physics journal vol 77) **3** 136-143 (in Russian)

[5] Anisimov A G 2010 *Investigation of the possibility of electric pulse sintering of nanostructured powder materials* **46 (2)** 135-139 (in Russian)

[6] Grigoryev E G 2009 Electric Current Pulse Welding of Titanium with 18-10 Stainless Steel *Advanced Materials Research* **83-86** 1251–1253
[7] Grigoryev E G 2011 High voltage electric pulse welding of titanium with stainless steel *Machines, Technologies, Materials* **8** 8–10
[8] Grigoryev E G and Olevsky E A 2012 Thermal processes during high-voltage electric discharge consolidation of powder materials *Scripta Materialia* **66** 662–665
[9] Nescoromniy S V and Strizhakov E L 2015 Method for obtaining compounds of heterogeneous conductive materials and a device for its implementation (Russian Patent No. 2551329 year 20.05.2015 Byul. No. 14.) (in Russian)
[10] Nescoromniy S V and Strizhakov E L 2010 Device for shock capacitor welding with magnetic pulse drive (Russian Patent No. 96515 year 10.08.2010) (in Russian)
[11] Strizhakov E and Nescoromniy S V 2014 Device for electric pulse sintering (Russian Patent No.136754 year 20.01.2014 Byul. No. 2.) (in Russian)
[12] Nescoromniy S V and Strizhakov E L 2008 Device for shock capacitor welding of rod parts with flat base (Russian Patent No. 70839 year 20.02.2008) (in Russian)
[13] Strigakov E L and Kem A Yu 2005 Technological features of obtaining of compound powder parts with usage innovative P/M processes for high performance metals -vacuum – thermal magneto – pulse pressing *16th International Plansee Seminar* (Reutte/Tirol, Austria **4** (14) 287-293
[14] Strizhakov E L and Neskoromny S V 2016 Development of discharge-pulsed equipment for applied studies of magnetic-pulsed welding processes *Welding International* **30** (10) 813 – 816
[15] Lemeshev S V and Nescoromniy S V 2015 Discharge – pulse pressing of composite materials *Bulletin of the don state technical University* **4** 76-81 (in Russian)
[16] Minko D V and Belyaev K V 2015 Pulsed electrophysical methods for obtaining composite materials and modified structures (Nanotechnologies and materials science in mechanical engineering, strengthening technologies and functional coatings SB. Tr. Based on the VII international scientific and practical conference. (*Innovations in mechanical engineering)* - BNTU Minsk) 337 (in Russian)
[17] Falcon E and Castaing B 2005 Electrical properties of granular matter: From "Branly effect" to intermittency (*in Powders & Grains*) 323 – 327
[18] Yusupov R Yu and Popov A 2007 Methods and technique of experimental studies of fast-flowing processes of MYOMAS (*Magnetic pulse processing of materials. Ways of improvement and development: Proceedings of the Inter-dunar. science.- tech. Conf. MYOM*-2007 Samara) 260-270 (in Russian)
[19] Medvedev Yu 2003 Formation of the powder material during electroplastic seal: the abstract *Diss. Cand. tech. Sciences* (SRSTU Novocherkassk) 18 (in Russian)
[20] Anisimov A G and Mali V I 2010 Investigation of the possibility of electric pulse sintering of nanostructured powder materials (FGV T. 46, No. 2) 135-139 (in Russian)
[21] Schütte P and Garcia Proc J 2009 Electro Discharge Sintering as a process for rapid compaction in PM-Technology *EURO PM* Copenhagen **3** 91-99
[22] Olevsky E A and Kandukuri S 2007 Consolidation enhancement in spark-plasma sintering: Impact of high heating rates *J. Applied Physics* **102** (11) 114913

IOP Conf. Series: Materials Science and Engineering 1005 (2020) 012020 doi:10.1088/1757-899X/1005/1/012020

The prospective technology of production of metal materials grains with extra high rate of solidification

E V Galkin and M V Zharov

Moscow Aviation Institute (National Research University)
Volokolamskoe highway 4, Moscow, 125993, Russia

E-mail: e.v.galkin@yandex.ru, MaximZharov@mail.ru

Abstract. This paper presents results of research of influence of extra high speeds of aluminum alloys grain solidification on the increase of mechanical strength characteristics of produced grains and semifinished products, which are made of grains by pressing. During the research it was determined that the most influential factor on metal grain and alloys solidification speed is the vapor film forming around the graining drop of liquid alloy. The vapor film forms due to transformation of the boundary layer into vapor state. The term 'vapor barrier' around graining object is implemented for the first time. The article offers the methods of removing the vapor barrier with an aim of a considerable increase of heat rejection intensity and, as a result, increase of grain solidification speed. It is determined that the only effective method of vapor barrier removal is a considerable increase of initial drop movement speed in the cooling liquid during the spinning of metal liquid alloy. The conducted experiment proved theoretical suppositions. The received semifinished products of grained aluminum alloys had a considerable upgrade of metal strength characteristics.

1. Introduction

The method of alloys peculiarities improvement by alloying has almost depleted its possibilities in respect to usual conditions of production of semifinished products of lingots. One of the prospective directions of metal and alloys quality improvement is granulating, which is based on usage of high rate of solidification of liquid alloy [1].

The essence of influence of rate of solidification on structure and peculiarities of multicomponent alloys results in creation on thin internal grain structure. In case of aluminum and low-alloyed alloys on its base, the crucial influence on the peculiarities is rended by the grain size. And in case of high alloyed aluminum alloys the peculiarities of the moulded state generally depend on the size of dendrite branches and the grade of development of dendrite porosity. The alteration of cooling-down speed of liquid phase before the beginning of crystallization and the speed of the mere process provide the fine crushing of internal grain driven by decrease of thickness of separate dendrite branches and the increase of quantity of these branches [2].

As the internal structure changes, the incrusting matters of second phase and internal micropores inevitably crush and allocate in a more uniformal way. The result of this process is a considerable inhance of mechanical peculiarities of the material.

Therefore, a purpose of this work is to explore the influence of extra high rate of solidification to aluminum alloys peculiarities improvement.

IOP Conf. Series: Materials Science and Engineering 1005 (2020) 012020 doi:10.1088/1757-899X/1005/1/012020

2. Materials and Experimental Methods

As the result of possibility of receiving a high rate of solidification, simplicity of machines and technology of production, the axifugal method of grain moulding received high popularity [3].

The method consists in the following: inside the rotatory sputter pour melted metal which under the effect of centrifugal force crushes into drops which are thrown through the holes in the cylindrical walls of sputter into the cooling-down liquid (water) [4].

It is important that during this method of grain moulding the cooling-down of drops which is occurring in water, provides reaching a considerably high speed of cooling-down of liquid-alloy drops, which accounts for 10^6 °C in a second for 1.0 mm diameter grains. Speaking about alloys of system Al-Cu-Mg the structure of grains, which are received by this method, is acutely crushed comparing with the structure of ingot of same composition. It is known that the presence of chrome, manganese, zirconium in high-duty alloys produce a positive effect on the complex of semiproducts mechanical peculiarities. It rises the temperature of alloy recrystallization, which facilitates the maintenance of effect of extrusion after the semiproducts thermohardening. High rate of solidification while grain moulding allows to receive abnormally supersaturated solid solutions. When the content of zirconium in the alloy is 0.4 – 0.8%, the initial intermetallics with zirconium are not present in the grain structure. The zirconium in this case is in the supersaturated solid solution. While heating before the thermal treatment and in the process of plastic deformation the supersaturated solid solution decays with the segregation of internal phase containing zirconium, which provokes lattice distortion. On a certain stage of the decay the hardening effect reaches the maximum value.

The conducted experiment aimed at increase of the grain solidification rate by the joint action of water eddy and cooling-down in the water solutions of substances which increase the water steam formation. 5-10% water solutions of alkali-metal chlorides (NaCl) and oxides of alkaline metals were used in the capacity of such substances. The transition from laminar to turbulent flow of the cooling-down liquid was produced with the application of swirlers. The conducted experiments showed the highest efficiency of such act in relation to increase of grain solidification rate.

The production of grains of alloys of system Al-Cu-Mg was accomplished by the axifugal method on a special laboratory facility [5].

The conceptual scheme of installation is represented on figure 1.

The installation for centrifugal spraying of grains consists of bell jar 1, which is at the same time bell jar for hardening tank, where the cooling liquid enters (service water) 2. The hardening tank has external rotative upper shell in the center 3 and internal rotative ladle 4, which receives the liquid metal alloy through a hole in the upper part 5. The rotation of external rotative upper shell 3 and internal rotative ladle 4 is synchronized. Due to centrifugal forces fusil metal is squeezed into the holes of internal ladle and passing through the holes of external moveable ladle, the metal gets into hardening medium.

It is known that when the speed of rotation of a water-logged body exceeds 3000 r/min, just as for the current facility, water is thrown aside due to the forming of airflow and water swirl, there appears air medium layer between the surface of rotative separating nozzle of the ladle and the surface of cooling liquid swirl. Due to the facility design it is possible to reduce to minimum the liquid alloy drop flight line in the air medium and, as a result, to keep the liquid alloy drops temperature at the moment of contact with cooling liquid equal to the temperature of liquid alloy in the casting mold, which favours the increase of graining speed. Cooling down in water, the grains get down to the bottom of external bell jar and discharge with service water through the inlet branch 6 in the bottom part of the facility. The electric engine 8 through the system of trucks and sockets provides the rotation of external nozzle of the ladle 3 and internal perforated ladle 4 with target speed.

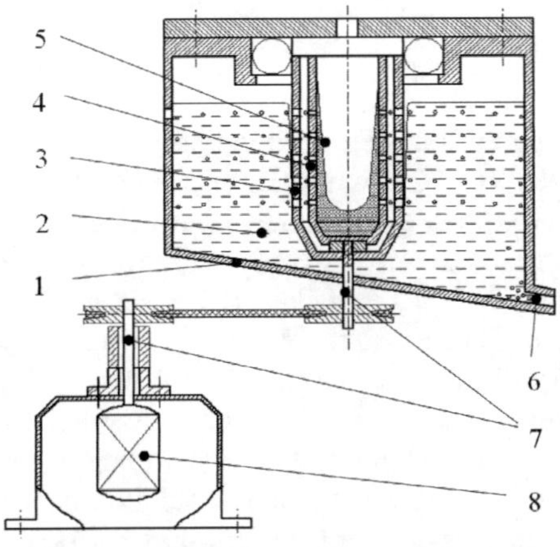

Figure 1. The conceptual scheme of facility for production of grains with extra high speeds of solidification [5].

The detailed description of facility concept is represented in [5]. The speed of liquid alloy ladle rotation is conceptually important, because it defines the requested travel speed of the dripping, which provides the vapor barrier removal around the crystalizing grain.

The rinsing and diffusion of grains into fractions were carried out according to traditional for metal power industry methodics. Nevertheless, while drying of grains the hot airflow drying on vibro-dryer conveyer was applied. It was done with an aim of decreasing the oxidation of grain surface due to the presence of water vapor while slow drying.

The grains had homogeneous grain finess and the diameter of about 1.0 mm. The prime yield after the grain moulding accounted for 90 %. The calculation showed the speed of grain cooling-down at this method of solidification. The received grains further on were prepared for metal forming. The heating was up to the temperature of 420 °C, the time of heating τ accounted for 8 hours. Then the grains were exposed to hot compressing with the production of a bar, a rolled section. The compressing was conducted in the laboratory fitting out on the fluid-actuated press by the force of 6000 kN with the extrusion ration λ =20÷30. The item endurance was conducted from the temperature of 480–500 °C and with the following artificial aging at the temperature of 190 °C during 8 hours.

3. Results and Discussion

The conducted research of structure and mechanical peculiarities of received granulated materials confirmed the theoretical assumptions. A considerable increase of alloy mechanical peculiarities was determined at the increase of grain solidification rate.

It is determined that the main difficulty on the way of increase of 1.0 mm grain crystallization speed is the forming of vapor barrier, which appears due to the convey of heat from crystallizing object to border layers of cooling liquid. As soon as the temperature of nearby water layers exceeds the temperature of boiling, appears the vapor barrier. Due to low heat conduction of vapor, the transition of heat from cooling down grain lowers sharply. The experiment proved that for the removal of vapor barrier it is required that the initial speed of liquid alloy drop was high enough at the moment of its entering into cooling water. The initial speed of grain movement is determined on the basis of lack of time for transition of required quantity of heat, which is needed for transforming the nearby water layers into vapor the volume of which does not exceed 1% of grain volume.

Overall, the speed of initial drop movement in the liquid alloy depends on metal liquid alloy physical peculiarities, ladle geometrical parameters and cooling liquid parameters. Also it was theoretically determined and experimentally proved that the needed initial drop speed depends directly on the drop radius, metal alloy thickness, ratio of drop heat dissipation to environment, temperature lapse (the difference of crystallizing drop temperature and cooling liquid temperature) and depends in inverse proportion on liquid alloy drop (grain) mass, volume vapor medium mass, cooling medium vaporization heat capacity, cooling medium heat capacity, temperature lapse between cooling medium vaporization temperature and its operational temperature.

At the determined initial grain speed the calculated grain cooling temperature achieves the speed of cooling of flakes which have a considerably smaller thickness than grains. The cooling speed of flakes (small flat sheets of abnormal form and small thickness) is under 10^8 °C/sec at their crystallization on copper backing (copper platforms) [1]. But because of their small thickness it is rather hard to conduct further pelleting and pressing of semifinished parts of produced flakes. It is a serious barrier on the way of industrial application of technology of production of moulded piece of flakes. It is determined that the sizes of particles may be regulated within determined limits during the centrifugal sputtering of liquid alloy. In case of sputtering and cooling down of liquid alloy by gas the speed of cooling down is lower when the particles size is equal, than at centrifugal diffusion into water. Nevertheless, the method of sputtering allows to receive the particles of a less size and due to this to come up to the speeds of cooling which are achieved at alloy diffusion into water. However, the industrial application of this technology faces the same problem as at pelleting and pressing of semifinished parts of flakes [3].

Overall the presented technology of aluminum alloy grains production by centrifugal separation applying the removal of vapor barrier is the unique graining technology which provides the grain target size with a high speed of their crystallization and makes its industrial application possible.

In the contex of alloys of system Al-Cu-Mg it is determined the increase of mechanical properties of semifinished rods received by the granulating technology. Table 1 shows mechanical properties of alloys AA2017 and AA2024 in the delivered state T1 (heat-treated and not naturally aged) of pressed samples, received by traditional technology of pressing bars of ingots (1), pressing of grains, received without the removal of vapor barrier at grain crystallization (2) and according to presented technology (3).

Table 1. Mechanical properties of pressed semifinished parts of aluminum alloys of system Al-Cu-Mg received by different technologies (1, 2 or 3).

Alloy	Production technology	Resistance to rupture, MPag	Yield offset, MPag	Percentage extension, %	Reference
AA2024	Pressing of ingot (1)	450	400	16	[6]
AA2024	Pressing of grains (2)	559–566	419–436	13.9–15.7	[1]
AA2024	Pressing of grains (3)	620–625	460–468	9.1–10.5	our data
AA2017	Pressing of ingot (1)	410	250	15	[6]
AA2017	Pressing of grains (2)	546	355	17.5	[1]
AA2017	Pressing of grains (3)	582–584	390–398	11.4–12.8	our data

4. Conclusion

The presented technology allows to produce high-strength aluminum grains of high quality , which are further used as semifinished raw part at production of parts by pressing, hot die forging or baking (after obligatory previous pelleting).

By this means, basing on the theoretical and instrumental research it is possible to evaluate good perspectives of the elaborated technology. The conducted technical-economic study showed that at a very insignificant cost increase of the semiproducts production and items of alloys of system Al-Cu-Mg it is possible to increase the strength characteristics of granulated materials by 15 – 18 % in relation to earlier observed high-strength aluminum alloys.

References

[1] Dobatkin V I and Yelagin V I 1981 *Granuliruemyie aluminievyie splavyi* (Moscow: Metallurgiya) p 176 (in Russian)
[2] Kolpashnikov A I and Efremov A V 1977 *Granulirovannyie materialy* (Moscow: Metallurgiya) p 240 (in Russian)
[3] Petrov A P, Yeremeev V V, Yeremeev N V, Krasnoborodko I O and Zlydnev I M 2017 *Scientific and technical journal Dvigatel*, **4** 4–8 (in Russian)
[4] Silin M B and Zharov M V 1998 *Method for production of metal grains* Patent No 2117556 (in Russian)
[5] Silin M B and Zharov M V 1999 *Facility for production of metal grains* Patent No 2125923 (in Russian)
[6] *Aviatsionnyie materialyi. Spravochnik. Part. 4. Aluminievyie i berilievyie splavyi. Part 1. Deformiruemyie aluminievyie splavyi i splavyi na osnove berillia.* Book No 1. 1982. (Moscow: ONTI VIAM) p 632 (in Russian).

IOP Conf. Series: Materials Science and Engineering 1005 (2020) 012021 doi:10.1088/1757-899X/1005/1/012021

Design of mechanical properties of structural materials for power plant equipment

V I Stashenko[1], O B Skvorcov[1,2] and O A Troickij[1]

[1] Mechanical Engineering Research Institute of the Russian Academy of Sciences (IMASH RAN), M. Kharitonyevskiy Pereulok 4, 101990 Moscow, Russia
[2] Scientific and Technical Center "Zavod Balansirovochnykh mashin", Varshavskoye highway 46, 115230, Moscow, Russia

E-mail: vis20-11@rambler.ru

Abstract. External impulse action on structural materials allows changing their mechanical properties. Such changes occur primarily in the surface layer of the material. One of the simplest methods of such an effect is to apply electrical impulses to metallic materials. Many different hypotheses have been proposed to explain the processes involved. In this paper, experimental data are presented that show that such hypotheses do not satisfactorily explain the observed phenomena. The results obtained show that the description of processes in the surface layer of a metal under the influence of an electric pulse should take into account the observed phenomena. This is the influence of the polarity of the applied electrical signal, the wave characteristics of the generated damped oscillatory processes, the synchronization of such processes with the moments of the beginning of the leading and trailing edges of the applied electrical pulse, the linear nature of the dependence of the amplitude of the vibration response on the amplitude of the applied pulse. The indicated features are manifested both in the region of elastic and plastic deformations of the metal. Such phenomena are observed for samples in the presence of additional static loading, and without it. The need for an adequate understanding of the processes occurring in this case is necessary for the effective application of methods related to the effect of electric pulses on metals. Such solutions are widely used to control the properties of materials in the processing of metals using cutting, drawing, when performing electric-pulse welding and the creation of non-destructive testing at the stages of production, operation and repair of various industrial equipment.

1. Introduction

The mechanical properties of structural materials used in conditions of radiation exposure vary significantly depending on the type of exposure, dose, material used and its processing. This applies to a large extent to materials with magnetic properties [1] and to such elements of electrical equipment as conductive elements. Excitation of vibrations and shock mechanical processes can be used in testing and non-destructive testing of metal conductive elements of power equipment [2]. By setting a high current density through such elements, it is possible to cause not only elastic, but also plastic deformations in them [3]. Elastic deformation in the material of the conductor when transmitting pulsed currents is manifested in the form of shock or damped oscillatory processes, which can be measured using high-frequency vibration sensors [4].

Significant elastic deformations (units and tens of microns) in the surface layers of conductors when pulsed currents are transmitted have a fusion on the properties of materials. Under such influences, a change in the phase composition and mechanical properties of the material is possible. These features are used, for example, during cold processing in the form of flattening or drawing using the electroplastic effect.

The manifestation of the mechanical action of high-energy electrical processes on the electrically conductive elements of powerful energy equipment is especially noticeable. Although the coefficient of conversion of electrical energy into mechanical energy is relatively small, the concentration of mechanical energy in the surface layers of the conductive structural elements can be achieved at relatively low current densities due to the skin effect. The impact on the surface layer of the material during metal processing improves its structure and uniform distribution of impurities and dislocations, in many cases it increases hardness and other mechanical parameters [5].

The vibroacoustic effect of the pulsed current passing through the conductors allows, in some cases, to replace the temperature and other effects that are difficult to implement in operating conditions of the equipment. In addition, the effect of high-density pulsed currents is accompanied by both mechanical and thermal effects, and the ratio between them can be changed by simple changes in not only the amplitude, but also the duty cycle of the pulses.

Such capabilities of testing [6] and monitoring of metal structural elements can be realized under the conditions of equipment operation, which is important, for example, if there are limitations associated with radiation safety.

Thus, the use of pulsed current influences allows combining the solution of both the problems of diagnosing the condition and processing materials directly on industrial equipment (for example, during repair or after it at the commissioning stage). Moreover, the force action of such pulsed currents allows you to change the properties of the material and reduce the influence of residual stresses [7].

2. Test procedure

The use of a diagram of the dependence of mechanical stresses on predetermined deformations (S-S diagrams) under the influence of electric pulses for studying ongoing processes is widely used. An example of such a dependence is presented in figure 1.

During the action of an electrical impulse, a discharge of the value of mechanical stress is observed. Based on the magnitude of such discharges and on changes in the general form of the dependence, conclusions are drawn about the nature of the physical processes occurring in the test sample.

This technique assumes the presence of additional static loading on the sample. The results obtained take into account only a relatively slow response to the effects of electric pulses due to the low speed of the measuring equipment used in such experiments. In addition, the dynamics of current changes during the action of an electric pulse are usually not taken into account.

To reduce the influence of these restrictions when performing measurements, we used a method for assessing the dependence of the acting mechanical forces and strains based on the measurement of axial and radial components of accelerations and measuring current changes through a sample with an accelerometer and a magnetic induction sensor, as shown in figure 2.

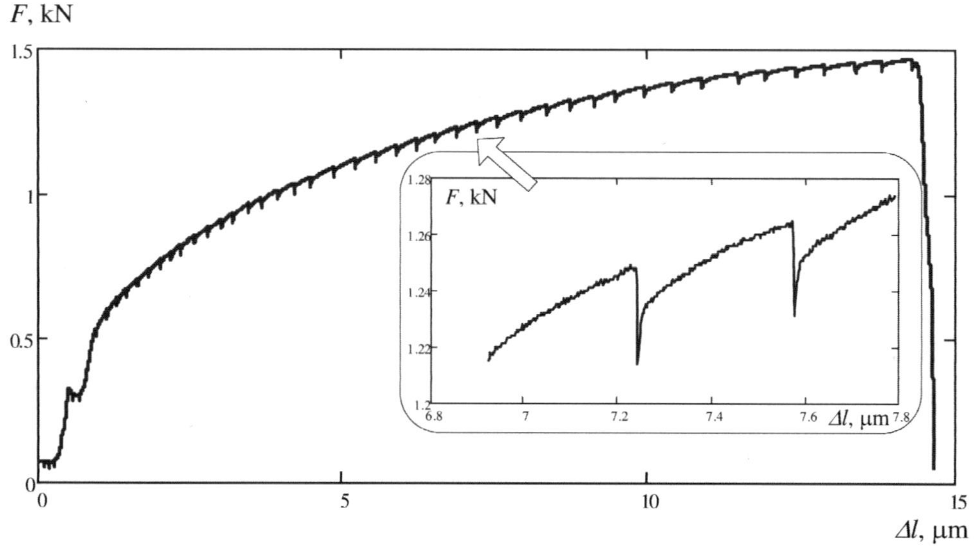

Figure 1. A typical diagram of the dependence of the tensile force on deformation under the influence of electric pulses. Rectangular stainless steel sample. An enlarged scale shows a part of the dependence for two adjacent current pulses.

This technique allows [8] to perform a synchronous analysis of the processes occurring at a high speed. This technique provides measurements, both under the influence of single unipolar electric impulses, and alternating pulses of opposite polarity. In this case, measurements can be carried out, as in the absence or presence of additional external static loads.

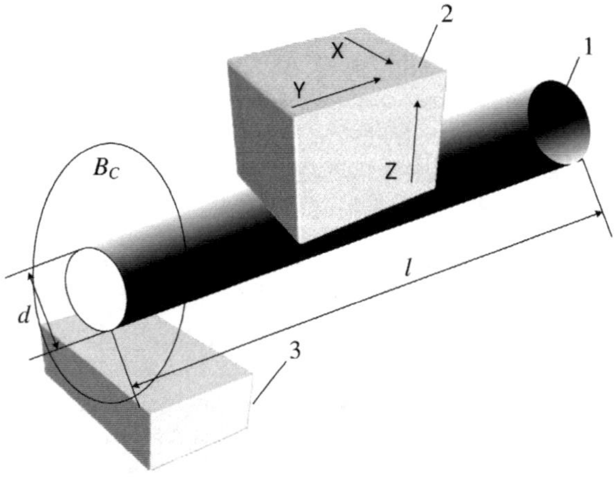

Figure 2. An example of an arrangement on a metal sample (1) of length l and diameter d of a multicomponent accelerometer (2) fixed with glue through an insulating gasket, as well as a non-contact magnetic induction sensor (3) under the sample.

3. Experimental data

When studying the vibrational response in samples of various metals (steel, aluminum, copper, brass, titanium, silver, gold) in the form of conductors of rectangular and circular cross section, vibrations in the axial and transverse directions are generated. It was found that the process of excitation of damped oscillations is associated with the formation of short-term processes similar to mechanical shocks at times corresponding to the beginning of the leading and trailing edges of the applied electric pulse [8]. Such damped oscillatory processes have opposite signs. The resulting oscillations are a superposition of such processes. At the initial moments of the fronts of the electric pulse, abrupt changes in magnetic induction are also observed, followed by a smooth change corresponding to the manifestation of self-induction and skin effect in the conductor.

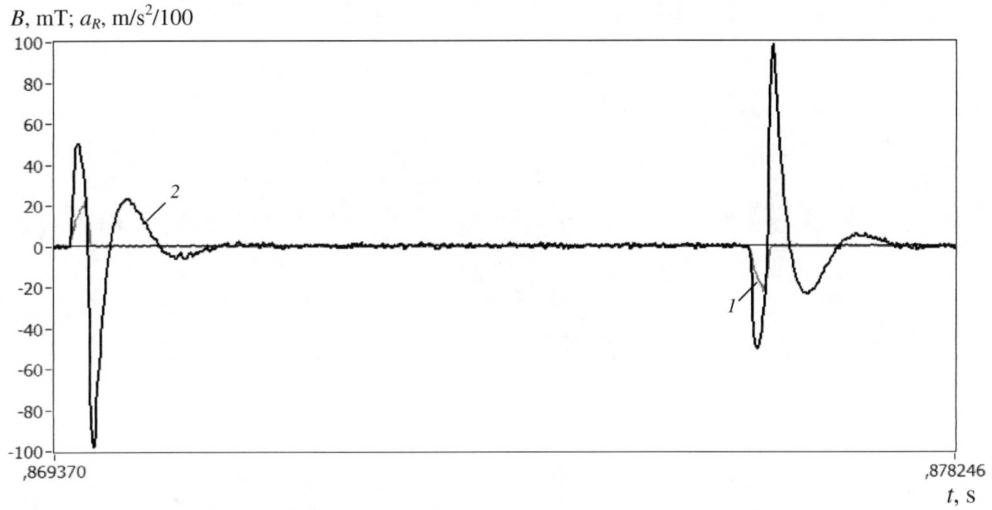

Figure 3. Diagrams of magnetic induction (1) and radial vibration (2) of the surface layer of a gold conductor with diameter of 1 mm from the action of neighboring electrical pulses of alternating polarity.

With a change in the direction of the current in the sample, not only a change in the sign of the magnetic induction pulse is observed, but also a change in the sign of the excited oscillations, as shown in figure 3. Moreover, the amplitude of the vibration response both from the results of measuring acceleration and for displacements linearly depends on the amplitudes of the current and magnetic induction, as shown in figure 4. Estimates of the displacement of the surface layers were obtained by double integration of the acceleration signal.

It is essential for understanding the processes that take place at the moment of the onset of the electric pulse that the current is close to zero and, therefore, the currents cannot be the cause of mechanical disturbances in the material of the sample that occur at these moments of time.

4. Discussion of the results

The presence of a connection with the polarity of the applied electrical impulses, a linear dependence of the amplitude of the mechanical response on the amplitude of the applied electrical impulse, the manifestation of such a vibrational response under small electrical influences and in the absence of external static loading for conductors made of various metals do not correspond to hypotheses explaining the mechanisms of deformation processes in conditions of electric pulse exposure.

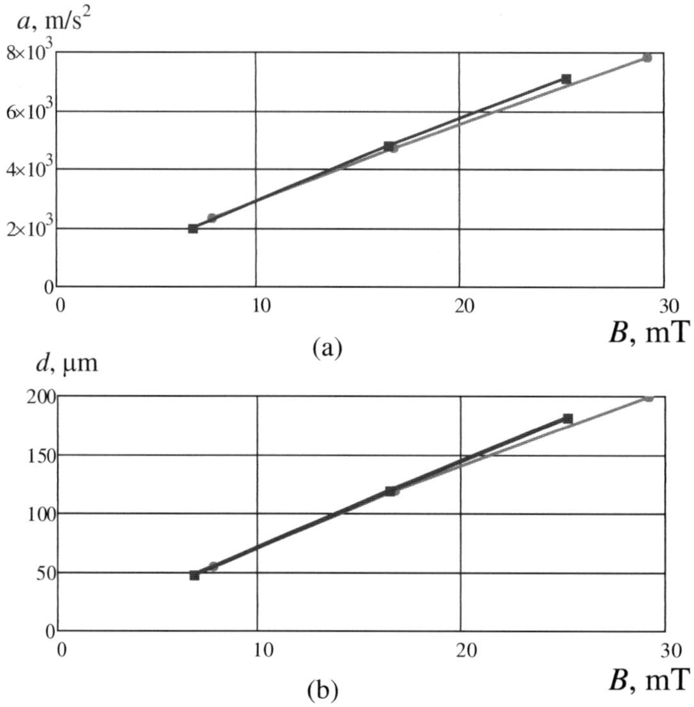

Figure 4. The dependence of the magnitude of the acceleration (a) and displacement (b) on the amplitude of the magnetic induction pulse. Sample gold diameter 1 mm.

Since the effect of the generation of vibro-acoustic waves in the conductor is present in the absence of external compression or tension for various materials, including in the absence of changes in the structure of the material, changes in this structure cannot be considered as the cause of the resulting deformations. A quantitative analysis of deformation processes under conditions when structural changes are observed in the sample material can show the possible contribution of other mechanisms to these processes. Such additional mechanisms may include the processes described in previously published hypotheses about the mechanisms of electroplasticity.

According to the results of the analysis of the data obtained, we can confirm the conclusion that there is no consistent description of the ongoing physical processes. Such a conclusion is contained both in the papers published in the last century [9, 10] and in the recent papers [11, 12] devoted to this issue.

5. Conclusion

For the effective use of this method of controlling the mechanical properties of the surface layer of metal products already used in products for various purposes, an understanding of the processes taking place is necessary. Such an understanding should be consistent with the obtained experimental data. Such experimental data, obtained both for simple conditions and in the absence of external loads, as well as the influence of temperature effects, contradict the existing hypotheses for describing the mechanisms of electropulse deformation. It is also necessary to conduct additional studies to determine the relationship and synergy of various previously proposed mechanisms under the conditions of creating both elastic and plastic deformations of metals.

References

[1] Petrov V V and Pupkov Yu A 2010 *Tests at INP. Radiation resistance of materials used to create magnetic systems of accelerators INP 2010-1* (Nonosibirsk) 1–33

[2] Troitsky O A, Stashenko V I and Skvortsov O B 2018 Vibrations of conductors during transmission of pulsed electric current and non-destructive testing *Engineering Journal: Science and Innovation* **3** (75) 1–16 (in Russian)

[3] Troickij O A, Skvorcov O B, Pravotorova E A and Stashenko V I 2018 Analysis of the relationships for the vibrational response to the excitation of vibro-acoustic processes in conductors from the action of a pulsed current *15th International School-Conference "New materials – Materials of innovative energy: development, characterization methods and application", KnE Materials Science* **2018** 611–620

[4] Troitskiy O A, Skvortsov O B and Stashenko V I 2017 Generation of mechanical vibrations in metal samples by the use of the pinch effect *IOP Conference Series: Materials Science and Engineering* **218** 1–6

[5] Troitskii O A, Skvortsov O B and Stashenko V I 2018 Measurements of vibrations produced by current pulses in elements of electrical equipment *Russian electrical engineering* **89** 3 143–146

[6] GOST 30630.0.0-99 1999 *Test methods for resistance to external factors affecting machines, devices and other technical products. General requirements* 1–41

[7] Lashchenko G I 2016 Technological possibilities of vibration processing of welded structures (Review) *Automatic welding* **7** (754) 28–34 (in Russian)

[8] Troitskii O A, Stashenko V I, Savenko V S, Skvortsov O B, Samuilov S D, Pravotorova E A and Tereshchuk V S 2019 *Impacts by current pulses and microwave studies on structural materials. Electrodynamic and electrochemical effects in conductors* Publishing house «Kim L A» 278 pp. (in Russian)

[9] Okazaki K, Kagawa M and Conrad H 1979 Additional results on the electroplastic effect in metals *Scripta metallurgica* **13** 277–280

[10] Bataronov I L 1999 Mechaninisms of electroplasticity *Sorovskii orazovat/ zurnal* 10 93–99 (in Russian)

[11] Lahiri A, Shanthraj P and Roters F 2019 Understanding the mechanisms of electroplasticity from a crystal plasticity perspective *Modelling Simul. Mater. Sci. Eng.* 27 085006 1–28

[12] Yu-chen Liu and Shih-kang Lin 2019 A critical review on the electromigration effect, the electroplastic effect, and perspectives on effect of electric current upon alloy phase stability *JOM* **71** 3094–3106 1-45 doi:10.1007/s11837-019-03661-y

IOP Conf. Series: Materials Science and Engineering 1005 (2020) 012022 doi:10.1088/1757-899X/1005/1/012022

Current state of studies of shock-wave action on the structure and properties of superconductors

B Mikhailov[1], A Mikhailova[1], V Nikulin[2], P Silin[2], I Borovitskaya[1], E Peregudova[2]

[1] Baykov Institute of Metallurgy and Materials Science, RAS, Leninskii pr. 49, Moscow, 119334 Russia
[2] Lebedev Physical Institute of RAS, Leninskii pr. 53, Moscow, 119333 Russia

E-mail: borismix@yandex.ru

Abstract: The article presents the results of the action of shock waves generated by the Plasma Focus setup on the structure and critical current Jc(B) of MgB_2 and Bi-2223-based superconductors. The structure of the tapes was studied in the transverse and longitudinal sections using different impact energy intensities (due to a change in the distance from the plasma anode from 20 to 50 mm for the same number of plasma pulses (n= 5)). The studies showed the possibility of increasing the critical current (Jc) up to 50% in transverse and longitudinal magnetic fields to 2-3 T after plasma impact due to changes in the microstructure of the superconducting phase, including: grain refinement, interlayer compaction, microdeformations and crystal lattice defects allowing to increase the number of pinning centers of magnetic vortices.

1. Introduction

The first experiments on the influence of shock waves (SW) with a pressure of 10^{10}-10^{11} Pa were used to increase the current carrying capacity of the commercial composite HTSC YBCO-tape (123) in [1,2]. To implement the submicrosecond shock wave action, the effect of the formation of SW arising from the interaction of high-speed cumulative plasma jet with a solid target is used. The basis of experimental studies of shock-wave action is the fact of the formation of point defects - vacancies and interstitial atoms (collective Frenkel pairs) at the front of SW as they pass through metallic and semiconductor materials. In this case, the concentration of emerging vacancies upon impact of plasma jet can be many times higher than the concentration of thermal vacancies in the initial material [3]. The formation of point defects in nickel under the influence of giant laser pulses and the effect of laser irradiation of niobium-based superconducting materials on the nature of critical temperature changes were shown in [4].

Taking into account the effect of the formation of point defects in metal and semiconductor materials at the front of shock waves [4,5] the laser influence on the temperature of the transition to the superconducting state (Tc) of niobium-based alloys was studied. These experiments showed a noticeable effect of shock waves influence on Tc. The observed effect was explained by the acceleration of diffusion processes due to the excess thermodynamically nonequilibrium defects arising during the passage of shock waves, as well as the role of dislocation interstitial loops. The results of these studies suggest that shock waves, depending on the power, can lead to significant changes in the superconducting characteristics and structure of HTSC tapes due to the creation of more equilibrium structural phase states and the formation of dislocation interstitial loops and vacancy pores. Studies of the physical processes occurring in superconductors in the state of contact with shock waves are in significant interest and open up wide possibilities for controlling the processes of structural, chemical, and phase transformations. Shock-wave action during the formation of micro and macrostructure can be selective and lead to different results. The action of a shock wave can cause high-speed crushing and mixing of particles and initiate ultrafast chemical reactions, contribute to the compaction of the microstructure, and lead to the appearance of various gradient defects.

The use of shock waves (SW) generated by various sources, including plasma pulses, dosed mechanical shocks, etc. attracts an interest to increase the superconducting parameters of a number of superconductors [4-15], including commercial tapes that are already ready for use.

The article presents some results of studies of shock effects produced by using a plasma focus type setup on the structure, grain morphology, phase composition, texture, and superconducting parameters Tc, Jc, Jc (B) of various superconducting compounds (wires and strips) based on Bi-2223 and MgB_2 [4-16].

2. Materials and Experimental Methods

The experiments were performed on the carved from industrial multicore wires and multilayer tapes samples based on Bi-2223, multifilament MgB_2 tapes and composite wires (D=0.9 mm) containing a stoichiometric mixture of magnesium and boron powders and on the tapes coated with superconducting powder, obtained using pressing and rolling.

To implement submicrosecond shock-wave action, a phenomenon of SW generation under the interaction of high-speed cumulative plasma jet with a solid target was used. A plasma jet was generated in a PF-4 plasma focus device (Lebedev Physical Institute, Russian Academy of Sciences) with a maximum energy stored in a capacitor bank of 4 kJ and a plasma jet velocity of 10^7 cm/s. The energy flux density in a jet impinging against the target surface reaches ~2 x 10^9 W/cm^2. The working chamber was filled with an inert gas (argon) at a pressure of 1.5Torr. The surface of the target (superconducting tape) was shielded from direct influence of plasma pulse with strips of metals with different electrical and thermal characteristics (Mo, Fe, Cu, Ti, and Al). The zone of SW action was 10 mm along the length of the tape, the distance from the anode to the tape surface was changed from 20 to 50 mm, and the number of plasma impacts was varied from 1 to 20. All plasma impacts were delivered on one side of the sample.

After the shock, the critical current value *Tc* was measured and the structure of superconducting interlayers was investigated. The critical current was measured in magnetic fields from 0.3 to 5.0 T at a temperature of 4.2 K.

X-ray diffraction phase analysis (XRD) was carried out by means of Rigaku Ultima IV at CuK_α-radiation using crystallographic database ICDD PDF-2 and software PDXL. Microstructure analysis was carried out using the optical microscope and scanning electron microscope EVA-40 (ZEIIS).

3. Experimental Results

Studies of the samples microstructure before and after plasma shock exposure allows you to get the visual picture of the changes that occur when applying SW. Figure 1 shows the microstructure of the cross section of multilayer Bi-2223 tapes in the initial state before impact (a) and after plasma SW (b).

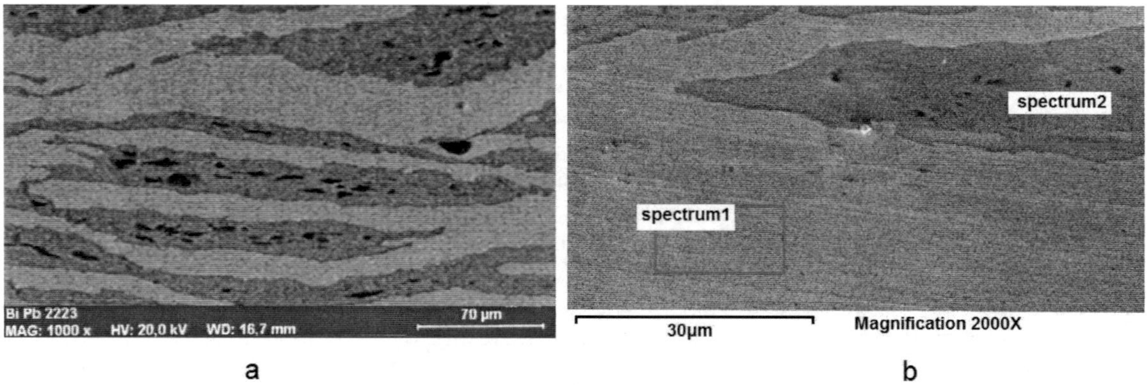

Figure 1. The cross section microstructure of Bi-2223 multilayer tapes in the initial state – before the impact (a) and after the plasma SW (b).

When comparing these structures, a significant compaction of the microstructure, the disappearance of the cracks, and decrease in the number of point pores in the microstructure core after SW exposure are noticeable.

Figure 2 shows with more magnification, the microstructure of the MgB_2 tape cross section in the initial state (a) and after the SW (b). After SW crushing and grinding of grains were observed, as well as the formation of a denser skeleton in MgB_2 core microstructure to compare with the sample before processing, which has a loose friable microstructure.

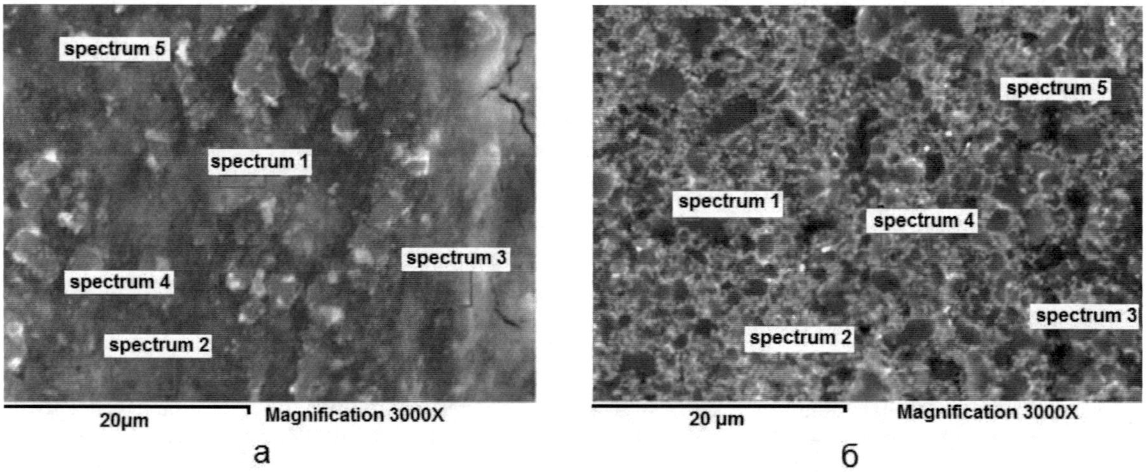

Figure 2. The cross section microstructure of MgB_2 core in the initial state (a) and after SW (b).

The change in the microstructure of the superconducting core after SW is also consistent with the XRD data. In a number of samples from MgB_2, Bi-2223, and Y-123 after SW, compared with the peaks in the X-ray diffraction patterns obtained before exposure, broadening of the X-ray peaks was established, and in some cases, with an increase of SW intensity, the formation of an amorphous phase has been established (5-10 wt.%)

The broadening of the peaks indicates the processes associated with a decrease in grain size and, consequently, size reduction of the coherent scattering regions (CSR) blocks. The broadening value is also depending of the microdeformations intensity in the structure of superconductors. On figure 3a. the X-ray diffraction pattern of the initial MgB_2 powder (Sp. Gr. P6 / mmm) is shown. The quantitative x-ray phase analysis was shown that the powder consists mainly of the MgB_2 compound - 94.4 wt% with an admixture of MgO ~ 5.6 wt% . The sizes of CSR (D) calculated in the PDXL

(Rigaku) program by the Williamson – Hall method for the initial powder were ~1000 A and the magnitude of microstresses - ε≈0%. Such values of these quantities are typical, as a rule, for annealed powders consisting of large crystallites. The MgB_2 lattice constants after refinement in the *Celref* program were - a = b = 3.0859(7)Å, c = 3, 5214(3)Å.

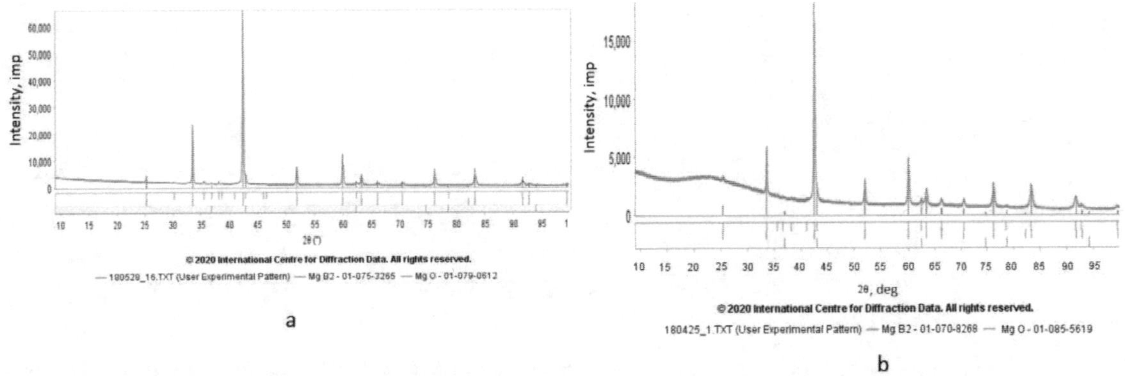

Figure 3. X-ray diffraction pattern of the initial MgB_2 powder (a) and X-ray diffraction pattern after rolling on aluminum substrate and SW action (b).

Further X-ray phase studies showed that the MgB_2 powders after treatment (Figure 3.b) remain almost single-phase and their phase composition does not change after processing by SW action of plasma - 95.6 wt% MgB_2 and 4.4 wt% MgO. However, the diffraction lines are noticeably broadened due to a change in the microstructure of the powder and grain grinding. The calculation of CSR (D) on the samples after SW action (5-10 strokes and 25-35 mm distance) showed that the CSR (D) decreases noticeably (to 230-300Å) from sample to sample. The values of microstrains (ε) are also slightly increase to 0.01-0.02%. In the range of 2θ (9°- 35°), a small amorphous halo appears (Figure.3b), which is apparently due to the small part of the powder goes into an amorphous state after a plasma shock. The refinement of MgB_2 lattice constants after SW showed that periods *a* and *b* after treatment are practically unchanged (3.0853(3)A). However, the *c* =3.5184 (3)A, which may be due to the appearance of defects in the crystal structure of MgB_2 under the influence of plasma shock waves. The formation of additional defects can contribute to an increase of the pinning force of magnetic vortices, which, according to the accepted theory of superconductivity upon application of a field, begin to attach to them and, as a result, prevent the penetration of an external magnetic field into the bulk of the superconductor.

Bi-2223-based samples were studied in more detail using the Rietveld structure refinement method in the MAUD software package. Figure 4 shows the x-ray spectrum of the initial Bi-2223 powder (Sp. Gr. A2aa), after refinement of the structure in the MAUD software package. In the refinement, background parameters, profile parameters, zero point position, peak asymmetry, lattice periods, the amount of each of the phases that make up the powder, texture parameters, atomic coordinate values, thermal corrections, and fill factors of atomic positions are taken into account.

The phase content in the initial sample was: Bi-2223 - 67.6 wt% and Bi-2212 - 32.4 wt%. The lattice periods of the main phase of Bi-2223: a = 5.419 (5) A, b = 5.418 (6) A, c = 37.195 (9) A. The sizes of the CSR blocks (D) and microstresses (ε) were specified. They respectively amounted to D ≈ 800 A, ε≈ 0.05% for the main phase of Bi-2223.

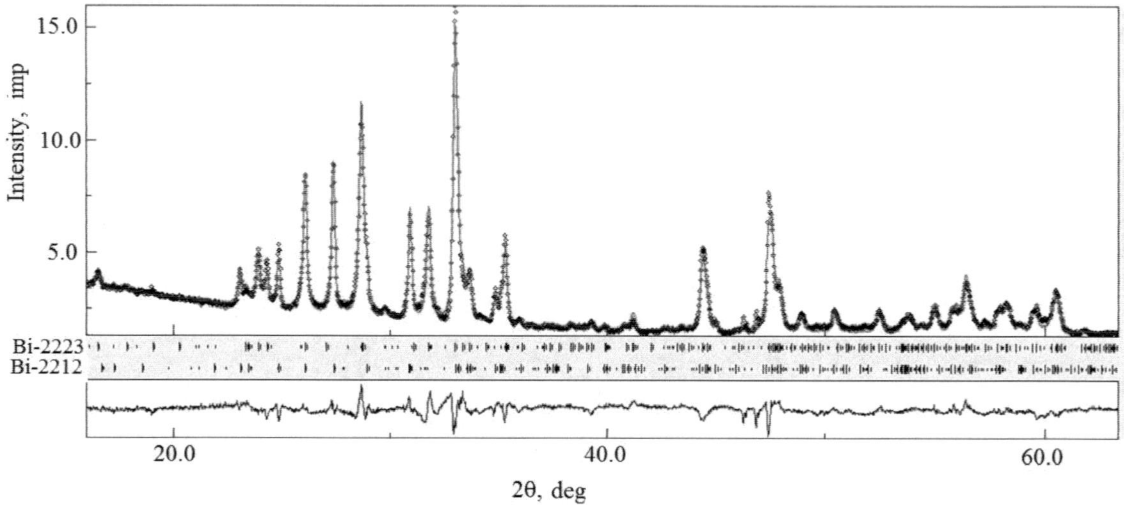

Figure 4. The results of the refinement of the structure according to the Rietveld method for the initial powder Bi-2223. The calculated spectrum is marked in red, black is the experimental one, the difference picture is shown under the figure. The divergence factors between the experimental and calculated profiles were: Rwp = 5.5, Rb = 4.3, Rexp = 1.9.

Next, we studied samples of coatings from HTSC ceramics based on Bi-2223 obtained by pressing and rolling on substrates made of copper and aluminum before and after treatment with a plasma focus. The distance between the anode and the sample was L = 35 mm, the number of strokes n = 5. Figure 5 shows the XRD results for such a sample.

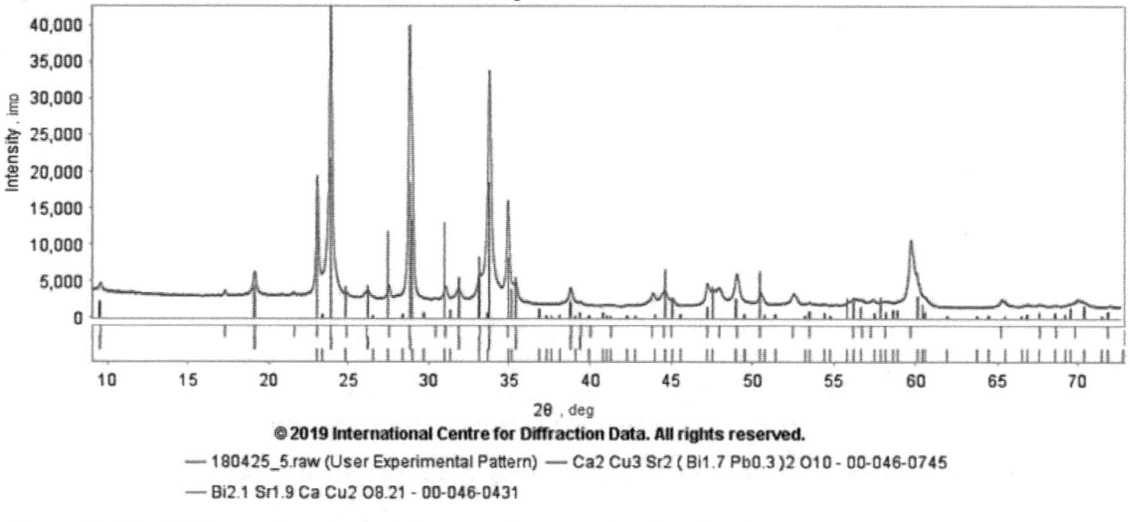

Figure 5. The XRD results of Bi-2223 ceramics sample after the shock wave action from the side of the plasma source.

The refinement by the Rietveld method showed that after SW decomposition of the superconducting Bi-2223 phase and degradation into simpler homologous phases was not observed. The number of Bi-2212 and Bi-2223 phases remains at the level of the initial sample: Bi-2223 - 68.2 wt% and Bi-2212 - 31.8 wt%.

This indicates, first of all, that the use of a plasma effect under the selected conditions (L = 35 mm, the number of strokes n = 5) made it possible to prevent overheating of the sample and its subsequent melting. When comparing the X-ray spectra for the initial powder and after SW action (Figure 5), it can be noted that the intensity of the lines of the sample after processing decreases, and the lines are very blurred, including reflections of the (001) type, which are characterized by a higher intensity due to the formation of texture rolling. At the same time, the preferential orientation of the grains is retained after SW processing. As in the case of MgB_2, the appearance of the peak broadening effect is noticeable, which indicates a strong grain refinement upon SW exposure, a decrease in the CSR (D), and the occurrence of microdeformations. The total visible effect after SW action of plasma exposure is expressed in obtaining more dense uniform microstructure of an ordered type. In macrovolume, annihilation of such defects as pores, cracks, loose fragments, and grain crushing can be observed. At the same time, microdeformations and microstresses appear in the structure of superconducting compounds.

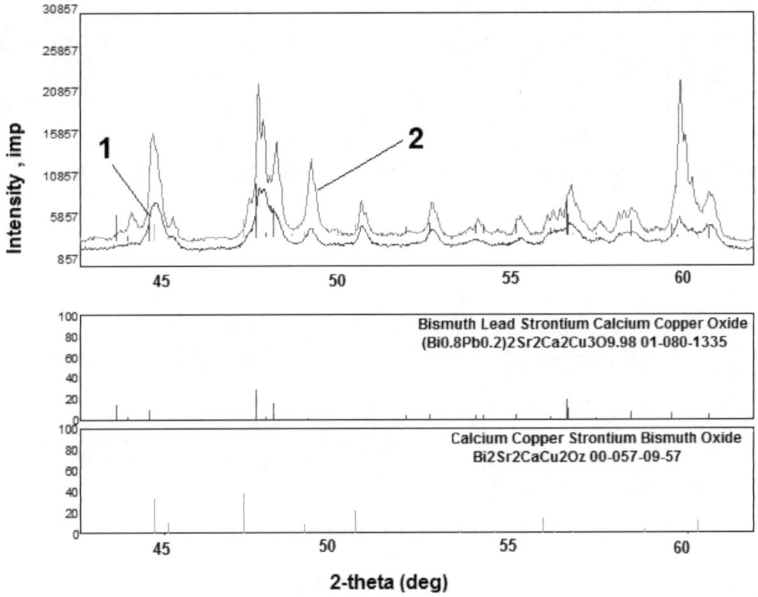

Figure 6. Comparison of fragments of powder diffraction spectra after SW action (1) and initial powder (2).

According to the calculations by the Rietveld method, the factors of divergence between the experimental and calculated profiles were: Rwp = 10.01, Rb = 6.54, Rexp = 1.98. These indicators are significantly higher than for the initial powder and after rolling.

This is due to the change in the microstructure and structure of the samples as a result of pressing, rolling and SW processing and, accordingly, a change of X-ray spectrum, including description by mathematical functions. The sizes of the CSR blocks and microdeformations for the sample after SW processing – for the Bi-2223 compound – were $D \approx 580A$, $\varepsilon \approx 0.05\%$. It can be seen that the size of the CSR is noticeably smaller than for the initial powder. The lattice periods were Bi-2223 - a = 5.386 (5) A, b = 5.423 (4) A, c = 37.188 (9) A. Changes in the lattice periods can be associated with the cyclical SW influence (in this case, n = 5) as a result of which the sample alternately experienced rapid heating and then gradually cooled. This could lead to disordering in the structure, as well as the appearance of defects and a change of lattice constants. To study such effects, it is necessary to eliminate the reflections of the second impurity phase (Bi-2212) from the spectrum, to exclude the contribution from the influence of texture and substructure, since these features of the sample very much distort the X-ray picture, and do not make it possible to refine a lot of parameters. As can be seen from the results of X-ray powder diffraction and microstructure studies on the examples of three superconducting compounds: MgB_2, Y-123 and Bi-2223, after processing by shock waves using a plasma source, grain size grinding, powder compaction, and better adhesion between grains were observed in all samples. The phase composition of the samples did not change, but the appearance of defects was observed in the structure, which expresses itself in a decrease of the lattice periods, and the size of the CSR blocks (D). Microdeformations became more noticeable. A certain amount of powder (1-10% - depending on the conditions of striking) acquired an amorphous structure. Similar changes in the structure and microstructure can explain an increasing of the critical current after SW treatment. Firstly, defects occur at the micro level, which leads to an increase in pinning force, and secondly, the powder is compacted by grinding and core compaction in the bulk of the cores themselves; moreover, grain boundaries are melted, which also improves contact between grains. Also, in the case of manufacturing samples using rolling and pressing, a texture is formed in the direction of the current flow, which remains after SW treatment. The material of the protective shield covering the sample from direct plasma exposure during SW processing also has a significant effect on the structure of the samples the critical current. The experiments with using screens of various metals with a thickness (100 μm) at the same distance from the PF anode (35 mm) showed the highest critical currents in the case of using titanium foil (850 A) in a magnetic field of 1.0 T at helium temperatures .

Depending on the exposure conditions: the distance between the tapes and the anode, and the number of pulses, as well as the type of protective gaskets used, all these effects can express themselves in different degree and lead to different results, basically in case of the critical current magnitude.

4. Conclusion

1. The possibility of the critical current increasing of superconducting tapes to 50% in 3–4 T magnetic fields due to shock-wave and temperature effects has been shown in a number of published works.
2. The increase in the critical current of the tapes can be explained by an increase in the density and hardness of superconducting layers, grain refinement, and an increase in the number of effective pinning centers.
3. The possibility of increasing the critical current MgB_2 tapes in the case of SW action using protective gaskets of various metals has been established.

Acknowledgements

This work was supported by the Russian Science Foundation, grant No. 16-12-10351 in the part of processing samples at the Plasma Focus installation and in the framework of the State task No. 075-00947-20-00 in the part of conducting structural studies.

References

[1] Aragwala P, Srivastava M P, Dheer P N et al 1999 *Physica C* **313** 87-92
[2] Murr L E, Niou C S, Jin S et al 1989 *Apll. Phys.Lett.* **55** (15) 1575–1577
[3] Mesokh Z I, Yanushkevich V A and Ivanov L I 1971 *FHOM* **(4)** 163–165
[4] Dekhtyar I Ya, Ivanov L I et al 1976 *Quantum Electronics* **3 (4)** 844–947
[5] Nikulin V Ya, Ivanov L I, Mikhailova G N, Mikhailov B P, Troitskii A V, Antonova L K, Borovitskaya I V, Gorshkov P V, Peregudova E N, Pokrovskii S V and Rudnev I A 2011 *Acta Technica* **56** T238-T244
[6] Antonova L Kh et al 2011 *Physics of Metals and Metallurgy* **111 (2)** 162–168
[7] Mikhailov B P, Ivanov L I, Shamrai V F, Nikulin V Ya, Mikhailova G N, Nizhankovsky V I, Rudnev I A and Gorshkov P V 2009 *Promising Materials* **6** 57–60
[8] Mikhailov B P, Ivanov L I, Borovitskaya I V, Krohin O N, Rudnev I A, Troitskii A V, Antonova L Kh, Nikulin V Ya, Mayorov A N and Pokrovsky S V 2012 *Reports of the Academy of Sciences, Technical Physics* **442 (5)** 614–616
[9] Mikhailov B P, Ivanov L I, Shamray V F, Nikulin V Ya, Mikhailova A B, Gaida D, Gilevsky A, Mayorov A, Rudnev I A and Pokrovsky S V 2011 *Books of articles of First National Conference on Applied Superconductivity NCPS I* "Kurchatov Institute" 23–29

[10] Mikhailov B P, Ivanov L I, Borovitskaya I V, Krohin O N, Rudnev I A, Troitskii A V, Antonova L Kh, Nikulin V Ya, Mayorov A N and Pokrovsky S V 2012 *Doklady of Physics* **57 (2)** 61–63
[11] Mikhailova G, Antonova L, Borovitskaya I, Krochin O, Majorov A, Mikhailov B and Nikulin V, Silin P and Troitskiy A 2013 *Physica Status Solidi* **10 (4)** 689–692
[12] Antonova L et al 2011 *The Physics of Metals and Metallography* **111 (2)** 158-164
[13] Kolokol'tsev V N, Mikhailov B P, Ivanov L I, Borovitskaya I V, Dorofeev Y A, Nikulin V Ya and Bondarenko G G 2012 *Inorganic materials: applied research* **3 (2)** 120–123
[14] Antonova L Kh, Borovitskaya I V and Gorshkov P V et al 2009 *Reports of the Academy of Sciences (Technical Physics)* **428 (4)** 471-473
[15] Mikhailov B P, Nikulin V Ya, Silin P V, Mikhailova A B, Mineev N A, Gaida D and Shamrai V F 2013 *Prospective materials* **10** 70–75

IOP Conf. Series: Materials Science and Engineering 1005 (2020) 012023 doi:10.1088/1757-899X/1005/1/012023

Creation of ultrathin niobium nitride films at temperatures less than 100 °C

B V Goncharov[1], B A Gurovich[1], K E Prikhodko[1, 2], M M Dementyeva[1], V L Stolyarov[1], E D Olshansky[1], A G Domantovsky[1], L V Kutuzov[1], E M Malieva[1] and A A Cherepanov[1]

[1] National Research Centre «Kurchatov Institute», sqr. Akademika Kurchatova 1, Moscow, 123182, Russia
[2] National Research Nuclear University (MEPhI), Kashirskoe highway 31, Moscow,115409, Russia

E-mail: goncharov_bv@nrcki.ru

Abstract. Thin NbN films were synthesized by a method of magnetron sputtering of solid Nb targets by nitrogen ions at temperatures less than 100 °C on substrates of the sapphire. The dependences of electrical resistance of films on the temperature in range 4.2 K – 300 K were measured. Volt-ampere characteristics and critical current density vs film thickness of films were made at a temperature of 4.2 K. The microstructure and film-depth chemical composition of thin films were investigated by analytical methods of the transmission electronic microscopy (HRTEM, EELS) on a cross-section samples made by the FIB technique. HRTEM and EELS study showed the $Nb_{0.84}N$ phase and poly-crystal structure of niobium nitride films. It was found that the critical current density falls down by an order of magnitude when the film thickness was less than 4 nanometers.

1. Introduction

Ultrathin superconducting NbN films synthesis at low temperatures is relevant for the tasks that are associated with the manufacturing of multilayer structures containing a large number of nanoscale functional elements.

High temperatures (800 –1000 °C) of the substrate during deposition are common for producing ultrathin NbN films, with superconducting properties at cryogenic temperature [1,2]. Thus creating a large number of functional nanoscale elements in every layer faces with the problem of maintaining properties. Deposition of each layer anneals all of the underlying. In this work we present the methods of manufacturing NbN films, produced at temperatures that are close to room temperature, with properties comparable to films, produced at temperatures of 800 °C and higher.

2. Materials and Experimental Methods

Thin films of NbN (thickness 3.75 – 7 nm) were produced using cathode and magnetron sputtering techniques at a substrate temperature of~100°C. Monocrystalline sapphire substrates were used.

The "microbridges" with the sizes of 20x20 μm, and nanowires of different geometry were used as samples for electrical measurement. Structures for electrical measurements were fabricated by photolithography and plasma-chemical etching in a manner similar to that described in [3]. Figure 1

IOP Conf. Series: Materials Science and Engineering 1005 (2020) 012023 doi:10.1088/1757-899X/1005/1/012023

shows the SEM image of the experimental samples used for electrical measurements. The metal Ni(50nm)/Pt(20nm) measuring contacts were made by cathode sputtering technique (Figure 1b).

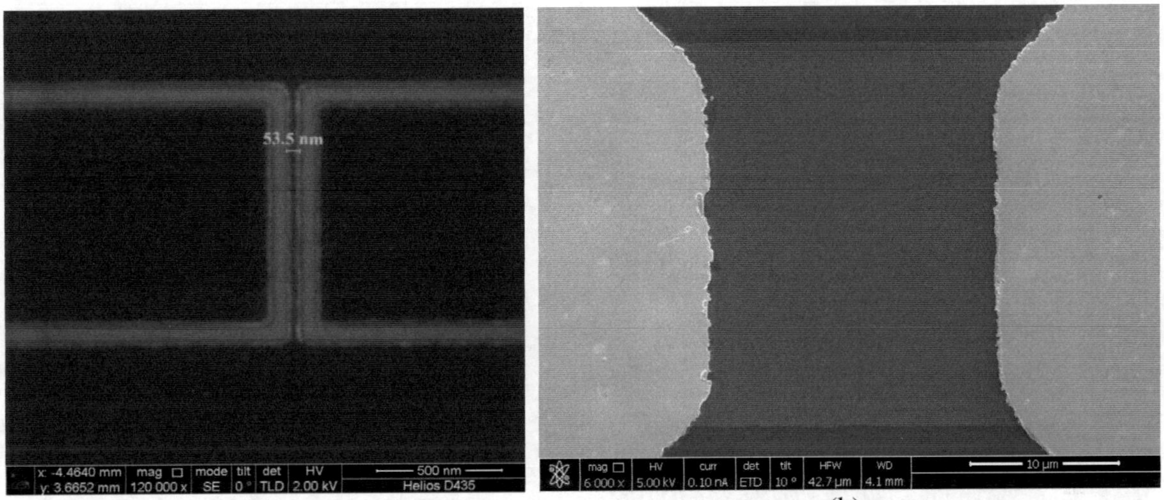

(a) (b)

Figure 1. (a) SEM image of nanowire and (b) NbN microbridge with Ni/Pt contacts.

Samples were cooled by placing them in a liquid helium. Measurements of temperature dependences of resistance were performed in a temperature interval from 4.2 K to 300 K at direct measuring current 100°μA. "4-wire" measure method was used. I(V) characteristic were measured at temperature 4.2 K using a Keithley-4200SCS parameter analyzer. Critical current densities were calculated using equation:

$$j_c = \frac{I_c}{S} \tag{1}$$

where I_c— direct critical current, S— sample cross-sectional area.

In this work we used transmission electron microscopy HRTEM for phase analysis and STEM coupled with electron energy loss spectroscopy (EELS) technique to characterize the elements depth distribution. Transmission electron microscope "Titan 80-300ST" operated at 200KV, equipped with a GIF-2001 energy loss spectrometer was used. The EELLS profiles were collected with an energy dispersion of 0.5°eV/ch and 2 s dwell time, convergence angle α was 10°mrad and collection angle β was 14.85°mrad. A cross-section samples NbN/Al_2O_3 were prepared by Focusing Ion Beam (FIB "Helios Nanolab 650") technique. Determination of phase composition in the initial samples was performed by the Fourier transform diffraction pattern obtained from the corresponding high-resolution transmission electron microscopy (HRTEM) image similar to the method used in the work [4].

3. Results and Discussion

It was shown by HRTEM investigations that the structural state of initial ultrathin NbN films were polycrystalline. Phase analysing data is demonstrated on figure 2(a). It was established that phase composition of individual grains corresponds to the R-3m rhombohedral $Nb_{0.84}N$ phase with cell parameters: a=b=2.985 Å, c=23.843 Å.

Quantitative analysing data were completely consistent with results of phase composition studies. Figure 2,b shows depth profile distribution of Nb, N, O atomic concentrations, calculated according to the EELS data. Increased nitrogen content in superconducting thin films were observed during magnetron sputtering at low temperatures. Phase analysis demonstrated that atomic concentration of nitrogen atoms in $Nb_{0.84}N$ phase corresponded to 55 at.%. According to the EELS analysis results this

IOP Conf. Series: Materials Science and Engineering 1005 (2020) 012023 doi:10.1088/1757-899X/1005/1/012023

the average nitrogen content was (64 ± 3) at.% (see fig.2,b). Increasing of oxygen concentration at the film surface was owing to oxidation film after sputtering. Thus the total film thickness was ~4 nm, and the effective superconductive $Nb_{0.84}N$ film thickness was ~ (3 – 3.5) nm.

Figure 3 shows sheet resistance $R_{sq.}(T)$ dependence for NbN films of different thickness. Value of superconducting transition temperature gets to a presumable interval (6 – 11) K.

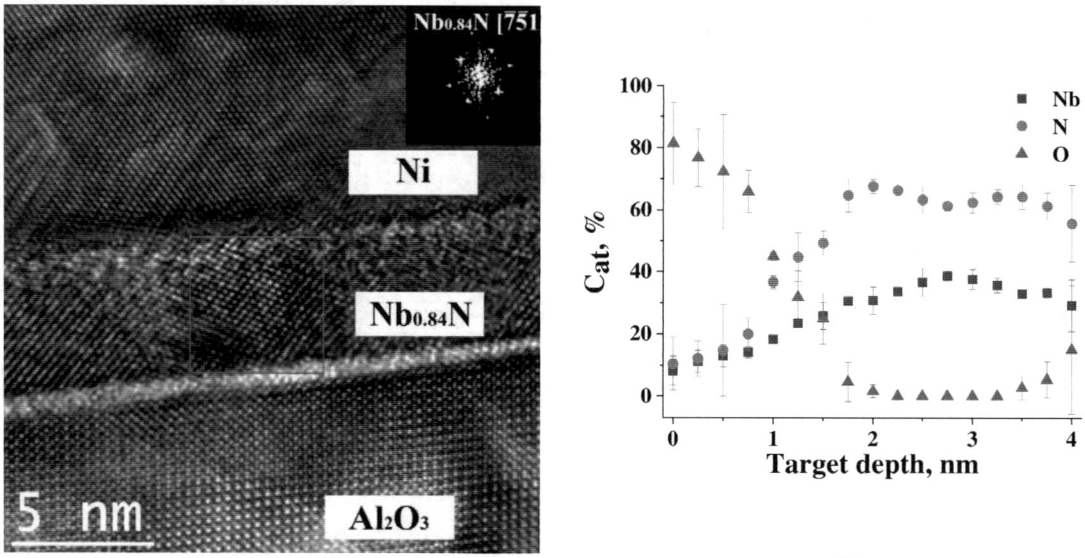

(a) (b)

Figure 2. HRTEM image of NbN thin film cross section sample, diffraction (insert) (a). Elements depth distribution profiles of NbN thin film in initial state, calculated according to the EELS data (b).

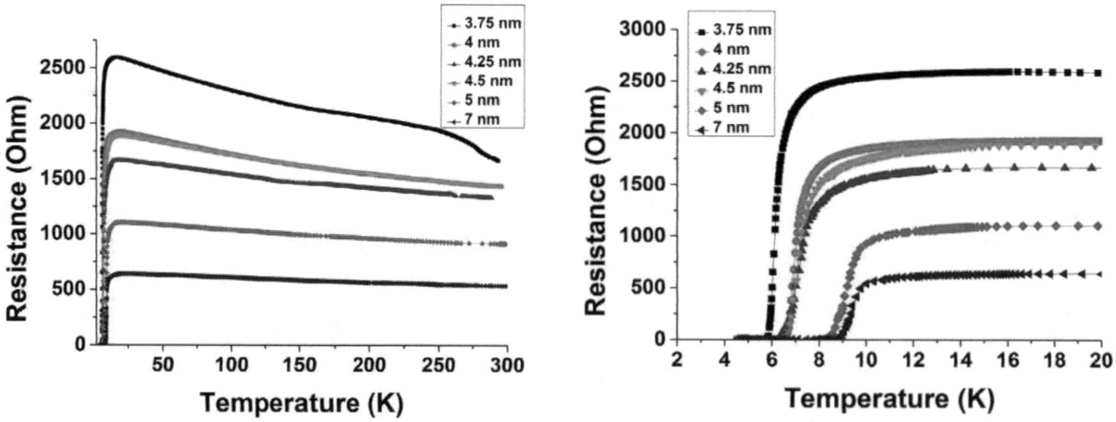

(a) (b)

Figure 3. Temperature dependence of electrical sheet resistance for NbN films of different thickness a) for 4 – 300 K , b) for 4 – 20 K.

The dependence of critical T_c temperature for the various thicknesses of NbN films on the film thickness is presented in Figure 4,a. It was shown that at thickness lower than 3.75 nm there were no superconductivity at 4.2 K liquid helium temperature. Current density vs film thickness is presented on figure 4,b and it is comparable with NbN films produces at high temperatures. When the film thickness is less than 4 nanometers, the critical current density falls by an order of magnitude.

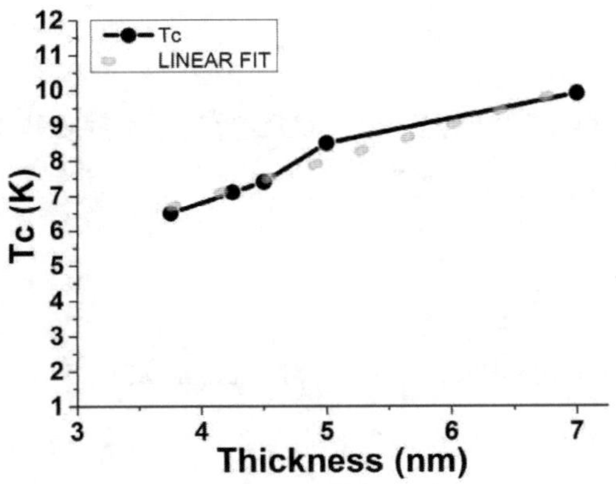

(a)

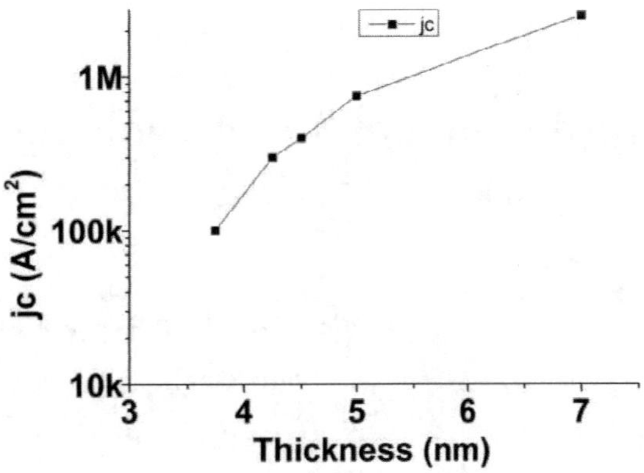

(b)

Figure 4. a) Dependence of T_c vs film thickness b) Critical current density calculated from V(I)s for different film thicknesses.

4. Conclusion

Thin NbN films were synthesized by the method of magnetron sputtering of solid Nb targets with nitrogen ions at room temperature on mono-crystal sapphire substrates. HRTEM and EELS investigations showed that niobium nitride film obtained at room substrate temperature was $Nb_{0.84}N$ phase and characterized by poly-crystal structure.

The measured dependences of electrical resistance of films on the temperature in the range 4.2K – 300K showed a typical trend for thin superconducting films. The received volt-ampere characteristics at a temperature of 4.2K allowed to calculate the values of critical current density for the films and also its dependence on the total thickness of NbN. The high density of critical current at a total thickness ~7 nanometers indicates wide range of application for NbN films produced at room temperature in the field of cryogenic nanotechnologies, and multilayer applications.

Acknowledgements

The work was supported by the National Research Centre “Kurchatov Institute” (№ 1359).

References

[1] Gurovich B A, Tarkhov M A, Prikhod’ko K E et al. 2014 Controlled modification of superconducting properties of NbN ultrathin films under composite ion beam irradiation *Nanotechnologies Russ.* **9** 386–390.

[2] Gurovich B A, Prikhodko K E, Tarkhov M A et al. 2015 Investigation of the Mechanism of Conductivity of NbN Thin Films, Modified Under Composite Ion Beam Irradiation *Micro Nanosyst.* **7** (3) 172–9

[3] Gurovich B A, Prihod’ko K E, Dement’eva M M et al. 2015 The use of ion irradiation for converting superconducting thin-film NbN into niobium oxide Nb_2O_5 *Nanotechnologies Russ.* **10** (7–8) 530–6

[4] Prikhodko K E, Gurovich B A, Dement’eva M M 2016 Study of phase transitions in NbN ultrathin films under composite ion beam irradiation *IOP Conf. Ser. Mater. Sci Eng.* **130** 1–5

Application of 3d technologies for microwave range electronic components creating

Kh Gadzhiev, S Gadzhieva and T Chelushkina

Dagestan state technical University, Imam Shamil Ave. 70, Makhachkala, 367000, Russia

E-mail: gadjiev.xad@mail.ru

Abstract. This paper presents the results of the study of microwave radio-electronic components electrophysical parameters manufactured on a 3D printer. Laser sintering technologies for metal semiconductor and ceramic powders were used as experimental research methods. The resonant properties of one-sided surfaces in the form of a Mobius sheet made on the 3D printer are analyzed. An explanation of a traveling and standing wave formation in the oscillatory contours of the microwave range of one-sided surfaces in the form of a Mobius sheet is presented. The possibility to create digital active phased antenna arrays based on one-sided surfaces in the form of a Mobius sheet is shown.

1. Introduction

In the radio-electronic industry are widely used 3D printers for the manufacture of metal or ceramic structural elements [1-3]. At the same time 3D technologies can be effectively used to create electronic components such as resistors inductors capacitors resonators and microwave antenna-feeder devices [4-6]. Existing 3D printers form structural elements as a rule from the same material [7]. It is possible to create electronic circuits in the form of metal structural elements on a ceramic basis, which limits the range of manufactured electronic components. It is advisable to manufacture electronic components in the same technological cycle from materials with different electrophysical characteristics. To solve this problem it is necessary to produce powder structures from different materials with similar sintering temperatures and weight and size characteristics. Alternate sintering of powders with different electrophysical properties in each layer will allow forming of electronic components with unique parameters that cannot be implemented using other technologies. For example it will be possible to produce digital active phased antenna arrays based on one-sided surfaces in the form of a Mobius sheet. Such antenna system will allow for direction finding of aircrafts in conditions of additive and multiplicative interference of natural and manufactured nature.

2. Materials and Experimental Methods

To produce digital active phased antenna arrays based on one-sided surfaces in the form of a Mobius sheet it is necessary to use a 3D printer with the ability to print products from different materials with electrophysical properties. Ceramics semiconductors and metal will be used as materials. Because of application of layer-by-layer laser sintering technology ceramic and metal powders can form a ceramic surface in the form of a one-sided surface of a Mobius sheet. This one-sided surface will be metallized in such a way that in the cross section it will have the capacitance properties in the form of two plates and a dielectric between them, and in the longitudinal section it will be two metal coils of the inductor.

IOP Conf. Series: Materials Science and Engineering 1005 (2020) 012024 doi:10.1088/1757-899X/1005/1/012024

As a result on the basis of capacitance and inductance a resonant oscillatory circuit of the microwave range is formed in which a traveling or standing wave can be formed depending on the geometric dimensions.

3. Results and Discussion

Figure 1 shows a 3D printer for printing products made of different materials by electrophysical properties. Instead of a single container with particles of powdered material for sintering with a movable bottom in the vertical direction are added several similar containers containing particles of powdered materials with different electrophysical properties which will be used in a given order to form each printing layer due to the horizontal movement of the mobile platform with containers over the working area for the product sintering.

After the formation of the particles sintering of one powdered material is completed all containers are simultaneously shifted to the side so that the container with particles of powdered material with other electrophysical properties can also apply the next layer to the working surface for sintering another material at the same level as the previously printed material. Thus in one layer it is possible simultaneously form several surfaces consisting of different materials with different parameters. At the end of the sintering process excess particles of powdered materials from this layer can simply be blown away or collected using a suction device to be placed into the same container from which these particles of powdered material were taken. To ensure high accuracy at the boundaries of the formed surfaces it is advisable to start sintering with those particles of powdered material that have the highest sintering temperature. It will avoid parasitic interference when sintering subsequent particles of the powdered material since the laser beam temperature for each subsequent particles of the powdered material will decrease and will not affect the parasitic melting at the boundaries of the previous material in this layer. After all the surfaces of one layer have been formed all free space in this layer must be filled with the most fusible material or other similar material that can be easily melted dissolved etc. when the product forming is finished or will be stored as a filling to increase water resistance strength or perform other auxiliary functions [8].

For the obtaining of ceramic semiconductor and metal powders for additive technologies can be used a plasmatron installation with ultrasound influence (figure 2). To obtain spherical powders and granules is used the cylindrical workpiece rotation around the horizontal axis and the workpiece end face is melted by a plasma jet of an arc plasmatron. The separation of molten particles from the end face edge under the action of centrifugal forces and the solidification of particles during flight in a gas environment will occur in contrast to known technical solutions under the influence of ultrasound on the separating molten particles. The standing wave inside the rotating cylindrical workpiece will determine not only size of the molten metal particle but the moment of its separation too. A similar effect will be synchronously exerted by a standing wave from an inert gas medium at the edge of the molten end face. Changing the frequency of ultrasonic vibrations allows to control the process of forming of certain size metal particles.

Thus, the method of forming metal semiconductor or ceramic powders for additive technologies in a plasmatron installation under the ultrasound influence solves the problem of powders forming with different electrophysical properties in the form of spherical particles of the same size for subsequent use in 3D printers.

On a 3D printer using metal and ceramic powders with different electrophysical properties can be made a microwave range resonator in the form of a one-sided Mobius surface for a digital active phased antenna array.

The ultra-high-frequency electromagnetic oscillations resonator in the form of a Mobius (figure 3) surface of metal-dielectric-metal structure is a closed microstrip structure with twisted plates for one 180-degree rotation. Figure 4 shows a Mobius ring section which shows the geometric dimensions as the cross-section radius r and is indicated half length of the inductor coil as l / 2. It is also shown how the metal plates 1 around the dielectric layer 2 forms the capacitive component of the contour, and how the outer plate passes into the inner one, and the inner one into the outer one.

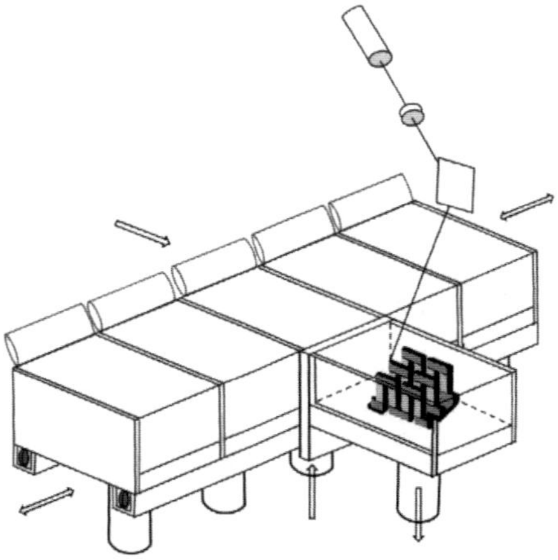

Figure 1. 3D printer diagram for printing products made of different by electrophysical properties materials.

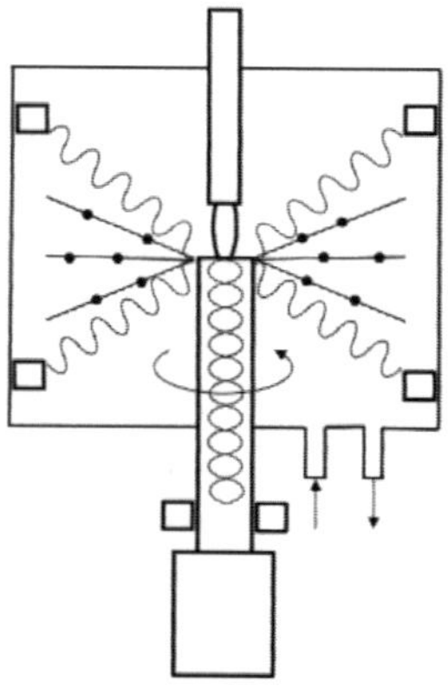

Figure 2. Method for forming powder structures with various electrophysical properties for additive technologies in a plasmatron installation under the influence of ultrasound.

In this case, two metal plates turn into one 1, which has the form of an inductance consisting of two turns closed on each other.

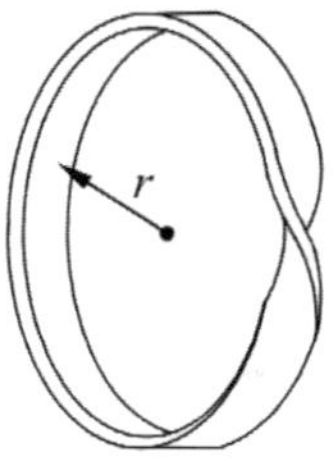

Figure 3. Resonator in the form of a closed surface of a Mobius strip.

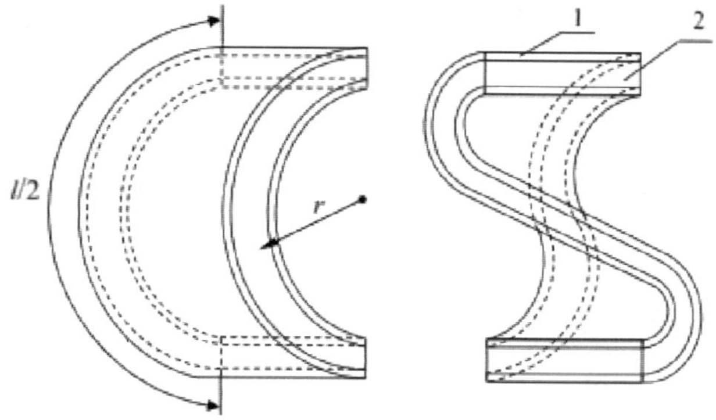

Figure 4. Section of the Mobius ring:
1 – metal plates;
2 – dielectric layer.

Figure 5 explains the formation of an oscillating circuit from a capacitance and two-turn inductance and shows the thickness h and width d of the capacitance indicating the metal plate 1 and the dielectric layer 2.

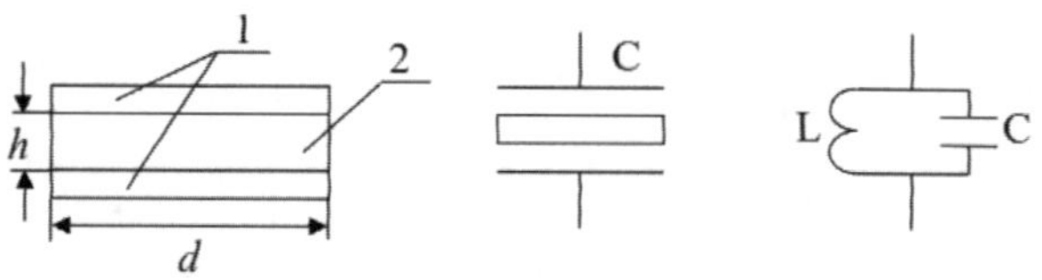

Figure 5. Formation of an oscillating circuit from a capacitance and two-turn inductance:
1 – metal plates;
2 – dielectric layer.

The resonator works as follows: in any section of the resonator there is a capacitive component in the form of two metal plates 1 on opposite sides of the dielectric surface 2. The capacitance value depends on the dielectric permittivity of the insulator its thickness (the distance between the metal plates) width of the plates and length, and the length of the plate corresponds to the length of one circuit turn. And inductance is a construction of a two-turn short-circuited solenoid that induces a magnetic field depending on the magnetic permeability of the dielectric size of the circuit cross-section and length of the metal plate. Inductance and capacitance together form an oscillating circuit. Depending on the geometrical dimensions of the circuit, it generates either a traveling wave (if a whole wave of oscillation is placed along the length of the plate) or standing wave (if half of the oscillation wave is placed along the length of the plate).

The change in inductance will be accompanied by changes in the capacitance of the circuit too since all the inductance parameters are also parameters of the capacitance. Some capacitance parameters can be changed without affecting the inductance value, so when changing the width or thickness of the dielectric and plates the inductance parameters will be saved but the capacitance parameters will change. Thus, it is possible to independently select the frequency of electromagnetic oscillations in the resonator by changing the capacitance parameters separately.

The calculation of the resonator electrophysical parameters (inductance, capacitance, and frequency) can be performed using the following formulas:

$$L = \mu_0 \mu \frac{N^2 \cdot S}{l}, \tag{1}$$

where L is the inductance, μ_0 is the magnetic constant, μ is the relative magnetic permeability, N is the number of turns, S is the cross-sectional area of the contour, and l is the length of the coil. If

$$S = \pi \cdot r^2, \quad l = 2 \cdot \pi \cdot r \Rightarrow r = \frac{l}{2\pi}, \tag{2}$$

then, after substitution, we get:

$$S = \frac{\pi \cdot l^2}{4\pi^2} = \frac{l^2}{4\pi}. \tag{3}$$

Substituting this value into the inductance formula for N=2 turns, we get:

$$L = \mu_0 \mu \frac{2^2 \cdot \left(l^2/4\pi\right)}{l} = \frac{\mu \cdot \mu_0 \cdot l}{\pi}. \tag{4}$$

The capacitance of the resonator's oscillating circuit is equal to:

$$C = \frac{\varepsilon \cdot \varepsilon_0 \cdot l \cdot d}{h}, \tag{5}$$

where C is the capacitance, ε is dielectric permeability, ε_0 is dielectric permittivity of vacuum, h is the dielectric thickness, l is the length of the plate, d is the width of the capacitance plate.

Substituting these values into the formula for calculating the frequency of the oscillating circuit

$$f = \frac{1}{2\pi\sqrt{LC}}, \tag{6}$$

get:

$$f = \frac{1}{2\pi\sqrt{\dfrac{\mu \cdot \mu_0 \cdot \varepsilon \cdot \varepsilon_0 \cdot l^2 \cdot d}{\pi \cdot h}}} = \frac{1}{2l}\sqrt{\frac{h}{\pi \cdot \mu \cdot \mu_0 \cdot \varepsilon \cdot \varepsilon_0 \cdot d}}. \tag{7}$$

Taking into account the formula for the electromagnetic oscillations speed $c = \lambda \cdot f$ is possible to find the wavelength:

$$\lambda = \frac{c}{f} = \frac{\dfrac{c}{1}}{\dfrac{1}{2 \cdot l}\sqrt{\dfrac{h}{\pi \cdot \mu \cdot \mu_0 \cdot \varepsilon \cdot \varepsilon_0 \cdot d}}} = 2c \cdot l\sqrt{\frac{\pi \cdot \mu \cdot \mu_0 \cdot \varepsilon \cdot \varepsilon_0 \cdot d}{h}}. \tag{8}$$

IOP Conf. Series: Materials Science and Engineering 1005 (2020) 012024 doi:10.1088/1757-899X/1005/1/012024

Thus, it is possible to determine the wavelength of the oscillation and compare it with the resonator size to select either the traveling wave mode or the standing wave mode.

For a digital active phased antenna array is important the pulse mode of operation with discrete values of information signals, which can be implemented in the form of a code-pulse modulator of ultra-high-frequency electromagnetic oscillations in the form of a multilayer Mobius surface with p-i-n diodes.

The modulator is a closed Mobius surface formed by a multilayer structure in the form of metal-dielectric-metal-p-i-n-diodes-metal-dielectric-metal, where each section of such a surface has a capacitive component formed by a rectangular dielectric on two opposite sides covered with metal plates. Through p-i-n diodes it is possible to connect or disable the additional plate in such a way that the capacitance will increase or decrease discretely (the inductance will remain unchanged). The closure in the form of a one-sided Mobius surface from these two metal plates forms one metal plate in the form of two turns of short-circuited inductance.

The modulator of ultra-high-frequency electromagnetic oscillations, in the form of a Mobius surface of the multilayer structure metal-dielectric-metal-p-i-n-diodes-metal-dielectric-metal is a closed multilayer structure with twisted plates for one 180-degree rotation. Figure 6 shows a section of the Mobius ring which shows how the metal plates 1 around the dielectric 2 form the capacitance component of the circuit, and how the outer plate passes into the inner one, and the inner one into the outer one. In this case two plates turn into one which has the form of an inductance consisting of two short-circuited turns. In addition it shows how p-i-n diodes allow to connect or disable the additional plate. The Mobius strip can be narrower (diodes are closed) or wider (diodes are open) depending on the polarity of the DC voltage applied to the diodes between the internal main and external additional plates.

Figure 7 explains the formation of an oscillating circuit from the capacitance and two-turn inductance using p-i-n diodes to change the resonant characteristics in the traveling and standing wave modes by connecting or disabling of additional plates to the capacitance in the modulator using p-i-n diodes.

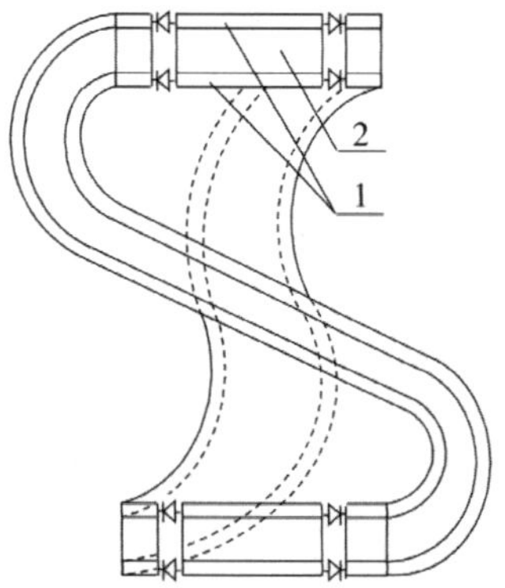

Figure 6. Modulator of ultra-high-frequency electromagnetic oscillations, in the form of a multi-layer structure Mobius surface: 1 – metal plates; 2 – dielectric layer.

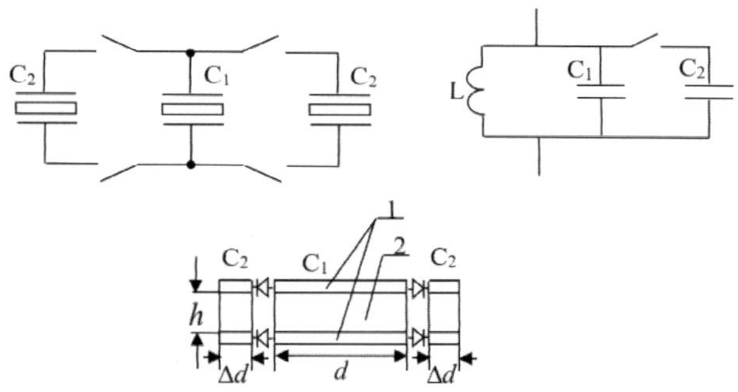

Figure 7. Formation of an oscillating circuit from a capacitance and a two-turn inductance using p-i-n diodes: 1 – metal linings; 2 – dielectric layer.

The modulator works as follows: in any section of the modulator there is a capacitive component in the form of two metal plates 1 on opposite sides of the dielectric surface 2. The value of the capacitor depends on the dielectric constant of the insulator, its thickness h (distance between the metal plates), plates width d and length l, and the length of casing corresponds to the length of one loop turn. With open p-i-n diodes, the plate will be wider by the value of Δd, and the capacitance is greater. Inductance is a two-coil short-circuited solenoid construction that induces a magnetic field depending on the magnetic permeability of the dielectric, size of the circuit cross-section and the metal plate length. Since the width of the plate does not affect the inductance the open or closed p-i-n diodes also do not affect the inductance value. Inductance and capacitance together form an oscillating circuit. When changing the width or thickness of the dielectric and plates the inductance parameters will be preserved, and the capacitance parameters will change. Thus it is possible independently select the frequency of electromagnetic oscillations in the modulator. By connecting of additional plate using p-i-n diodes the capacitance can be discretely changed, which in its turn results in a change of the circuit resonant frequency and allows to implement code-pulse frequency modulation [9].

Oscillatory contours of the microwave range in the form of multilayer Mobius surfaces can be used in digital active phased antenna arrays for communication and radar systems [10].

4. Conclusions

As a result with the help of the developed construction digital active phased antenna arrays can be manufactured on the basis of one-sided surfaces in the form of a Mobius sheet, consisting of conductive metal surfaces and dielectric ceramics.

Errors when printing on a 3D printer are comparable to tens of microns, which is many orders of magnitude less than the wavelength of the microwave oscillation of the centimeter and millimeter range, so such defects will not practically affect the accuracy of digital active phased antenna arrays, for which details less than a quarter of the wavelength are indistinguishable.

The microwave range resonator can be implemented using only additive technologies, since the use of semiconductor powders with different types of conductivity allows for the implementation of adjustable microwave range resonators. The microwave resonator of electromagnetic oscillations in the form of a Mobius surface of the metal-dielectric-metal structure can be used in communication and radar systems. It is advisable to use a resonator in the input circuits of digital active phased antenna arrays. An important advantage is that the electromagnetic oscillations of such circuit can be perceived by a similar circuit, rather than traditional microwave range resonators, that will protect the transmitting and receiving information from interference and unauthorized access. Also in the low-frequency range it is possible to produce thermoelectric devices, semiconductor components, etc.

References

[1] Nikiforov S O, Markhadaev B E, Nikiforov B S and Sholokhov E S 2016 *Bulletin of the Buryat scientific center of the Russian Academy of Sciences Siberian branch* **4** 156–163 (in Russian)
[2] Drovosekova T I Voronin A Yu and Zverko E K 2018 *Forum of young scientists* **4** 452–460 (in Russian)
[3] Yakushin K E Romanova E B and Kirillov G R 2018 *Fundamental and applied researches in the modern world* **22** 86–90 (in Russian)
[4] Dobychina E M Snastin M V and Solod A G 2018 *Telecommunication* **12** 68-73 (in Russian)
[5] Slatin V V 2017 *Aviation systems* **7** 3–9 (in Russian)
[6] Quaranta P 2016 *Military technolodgy* **9** 86–89 (in Russian)
[7] Vitsukaev AV Pavlovich O V Tsarkova Yu M and Maslennikova A A 2018 *Electronics and Super high frequency microelectronics* **1** 58–62 (in Russian)
[8] Ismailov T A Gadzhiev H M Shkurko A S Chelushkina T A and Kablov E N 2019 3D printer for printing products consisting of materials with different electrophysical properties Patent RF no. 2702019
[9] Ismailov T A Gadzhiev H M Chelushkina T A Shkurko A S and Magomedova P A 2015 *Bulletin of the Dagestan State Technical University* **37** 44–49
[10] Ismailov T A Gadzhiev H M Kryachko A S Chelushkin D A and Shkurko A S 2017 Code-pulse modulator of ultrahigh-frequency electromagnetic oscillations in the form of a multilayer Mobius surface with p-i-n diodes Patent RF No. 2616440

IOP Conf. Series: Materials Science and Engineering 1005 (2020) 012025 doi:10.1088/1757-899X/1005/1/012025

Slot-die printed Co_3O_4 nanolayers

V Koshlakova and D Muratov

National University of Science and Technology "MISIS"
Leninskiy prospect 4, Moscow,119049, Russia

E-mail: VikaKoshlakova@yandex.ru

Abstract. In this paper, we report on the possibility of obtaining uniform cobalt oxide nanolayers by slot-die printing of the dispersion onto conductive substrates. Dispersions for printing were obtained as a result of ultrasonic treatment of a mixture of cobalt oxide nanopowder with water. Nanopowder was synthesized by wet chemistry followed by heat treatment at a temperature of 200 °C. The resulting powder has platelet morphology. The chemical composition, uniformity, structure and size of the powder and nanolayers of cobalt oxide were studied using X-ray diffraction analysis (XRD), scanning electron microscopy (SEM), energy-dispersive X-ray microanalysis (EDX). The proposed easy and cost-effective method of slot-die printing from dispersion can be prospectively used to the production of cobalt oxide nanolayers for development of thin-film electronics and solar cells.

1. Introduction

In recent years, a huge part of scientific research is aimed at improving the environmental situation in the world. One of the main dangerous factors for the environment is to generate electricity by burning fossil fuels. Finding new and developing existing alternative energy sources will reduce the amount of harm to nature. One of the best options for alternative energy sources is the use of photoelectric technology to convert solar radiation into electricity. At the same time, when searching for alternative fossil fuels, you must consider the cost of new technologies. Based on these criteria, the most promising alternative energy source is perovskite solar element, since it has a relatively low cost and high development rates. However, the instability of perovskite solar element is a significant disadvantage and limits its use on industrial scale.

Usually perovskite solar element is collected from two electrodes between which are active layer perovskite and layers for transfer of electrons and holes. Improvement and development of new transport layers is the main option of solving instability of perovskite solar element. In addition, the use of the appropriate hole-transport layer can improve the efficiency of the transformation of sunlight into electrical energy. The main criterion for selecting material for the hole-transport layer is the value of its prohibited zone, which should reduce the barrier between perovskite layers anode [1, 2].

One of the promising materials for the hole-transport layer is cobalt oxide. This material has a suitable prohibited zone (~ 5.3 eV), high conductivity and heat resistance, environmental and environmental resistance [3-5]. Currently existing methods of obtaining holes-transport layers based on cobalt oxide cannot be Currently existing methods of obtaining holes-transport layers based on cobalt oxide cannot be implemented on an industrial scale, as they are costly or require large energy costs. In this regard, the purpose of this scientific work is to develop the method of obtaining Co_3O_4 nanolayers for use in perovskite solar elements, which will solve the problems described above.

2. Materials and Experimental Methods

The production of cobalt oxide nanolayers occurs consistently in two stages: synthesis of nanoparticles and applying dispersion on substrate. First, prepare a solution of 1M $CoCl_2 \cdot 6H_2O$ (cobalt chloride) and 2M NaOH (sodium hydroxide) for obtaining $Co(OH)_2$ (cobalt hydroxide). Mixing of solutions occurs with constant mixing, using magnetic bag. The process is completed when pH=10 is reached. To determine the pH solution is used pH-meter. The resulting blue-green sediment is filtered and washed with DI - water at least three times. After which the precipitation is sent to drying in the coupler furnace all night at 80°C. the obtained hydroxide cobalt is crushed and subjected to thermal treatment in the coupling furnace at 200°C within 2 hours. Nanopowder Co_3O_4 mixed with DI - water (H_2O) with 5 mg/ml concentrations and affected ultrasound for 1 min with subsequent centrifugation (5 min, 3500 rpm/min). The obtained dispersions were applied to pre-purified substrates of glass with ITO coating by the slot-die printing method. The printing temperature is 80ºC. One pass of the printed head is applied one layer of cobalt oxide. Substrates were cleaned with consistent ultrasound effect within 15 minutes in three different environments. X-ray diffraction phase analysis (XRD) was carried out by means of Diphre 401 at CrKα-radiation using Crystal-morphic database JCPDS. The definition of the chemical company of the nanopowder formed during temperature processing was carried out using the scanning electron microscope (SEM) "Vega 3SB" (Tescan) with an energy-dispersive X-ray microanalysis spectrometer (EDX) INCA x-act (Oxford Instruments). The Ramman - spectroscopy was produced using the DXR Raman -spectrometer "Thermo Fisher Science" (USA). The control of the size of the cobalt oxide particles was carried out using the analyzer of the size of particles Zetasizer Nano ZS.

3. Results and Discussion

Figure 1 shows X-Ray of synthesized powders before and after thermal treatment to confirm the necessary phase. Registered diffraction peaks of the original powder are indexed on hexagonal symmetry phase $Co(OH)_2$ (JCPDS file № 30-0443). The X-Ray powder after thermal treatment corresponds exclusively to the structure of cubic spinel Co3O4 (JCPDS file № 42-1467, spatial group: Fd-3m). Diffraction peaks at 2Θ = 19º, 31º, 36º, 38º, 44º, 55º, 59º, 65º, 68º, 74º, 78º grille plane (111), (220), (311), (222), (400), (331), (422), (511), (440), (531), (442), (620) accordingly. X-Ray confirm for full transformation $Co(OH)_2$ in Co_3O_4, without any additional phase.

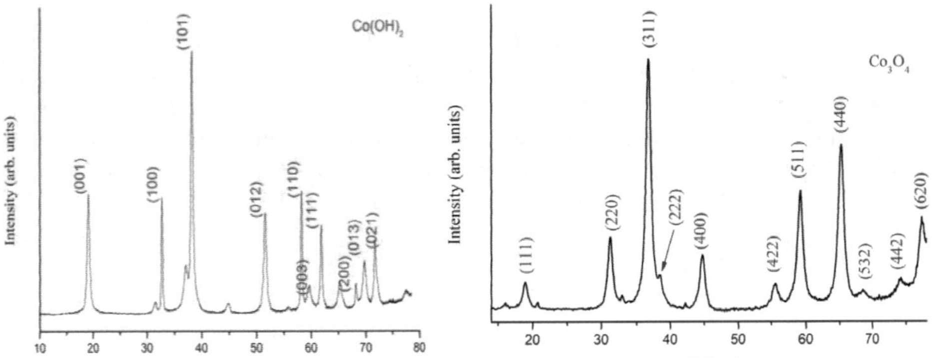

Figure 1. XRD of synthesized powders before and after heat treatment.

To determine the optimal method of preparation of dispersion with the smallest size distribution, various methods of obtaining dispersion (Table 1) were considered. The results of the size distribution of dispersion particles showed the tendency of cobalt oxide nanoparticles to coagulation, therefore to obtain a qualitative nanolayers of Co_3O_4 is necessary, preparation of dispersion directly before applying.

Based on the analysis of the size distribution of Co_3O_4, the method of obtaining dispersion using centrifugation was selected. Particles are characterized by a narrow size distribution with a average size of 58.7 ± 30.1 nm. The largest particle size does not exceed 120 nm.

Table 1. Size distribution of particles under addition influence.

№№	Kind of influence	Size distribution
1	-	87.5 ± 20.1
2	Filtration	68.1 ± 30.1
3	Centrifugation	58.7 ± 30.1
4	US-barh	122.4 ± 20.8

Figure 2 shows microphotography with scanning electronic microscope for cobalt oxide powder obtained at 200 °C. The structure of the Co_3O_4 has a plate shape. This type of structure is characteristic for multi-phase Co_3O_4. The structure strongly affects the possibility of applying dispersion methods from the powder data.

Figure 2. Microphotography of Co_3O_4 nanopowder.

Raman - spectra were recorded to confirm the presence and phase composition of the Co_3O_4 nanolayers (Figure 3). Five Raman peaks are observed on the spectra at 194.2, 479.3, 521.2, 617, 688.6 cm^{-1}, which correspond to phonon oscillations of single-phase Co_3O_4. When the number of layers increases, the intensity of the peaks decreases, which indicates a decrease in the size of the nanoparticles in the places where the spectra are removed.

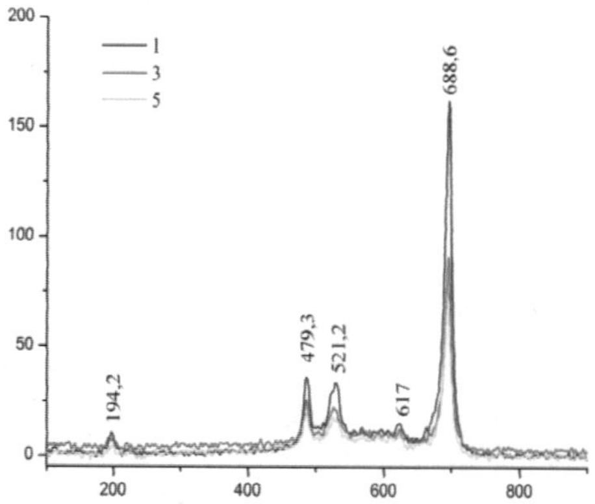

Figure 3. Raman spectrum of Co_3O_4 nanolayers.

Based on the transmission spectra, a significant decrease in the average transmission value was found with an increase in the number of layers. Samples obtained by slot-die printing with five Co_3O_4 nanolayers have a transparency of less than 60% over the entire wavelength range. Such a low value may be due to surface quality, which may cause light scattering. This value of the film transmission is unsatisfactory, which may lead to a decrease in incident light and a decrease in the overall efficiency of the device. A sample with one nanolayer has a maximum transmittance value and may have an uneven substrate coating. It was decided that a sample with three nanolayers is optimal in terms of uniformity of the substrate coating and transmission. For samples with one and three nanolayers in the wavelength range from 500 to 550 nm, there is a reduction in transmittance of up to 70%, which is typical for Co_3O_4.

On microphotography from an optical microscope (Figure 4), it can be observed that the sample obtained by slot-die printing with one nanolayer has rare small clusters of nanoparticles. With increasing layers, the number of clusters increases, and their distribution becomes more uniform, which with a large number of layers can lead to the formation of a uniform dense film. These clusters with a sufficiently large size reduce sample transmission and may cause an increase in the device's resistance, which will lead to poor conductivity of the hole-transport layer.

Figure 4. Microphotography of Co_3O_4 1 - nanolayers (left) and 5 - nanolayers (right).

Figure 5 shows microphotography from a scanning electron microscope for samples with five nanolayers. These nanolayers have a mesoporous morphology and a small number of agglomerated spherical nanoparticles of a larger size. Microphotography show that nanoparticles cover the entire surface of the substrate without forming visible holes.

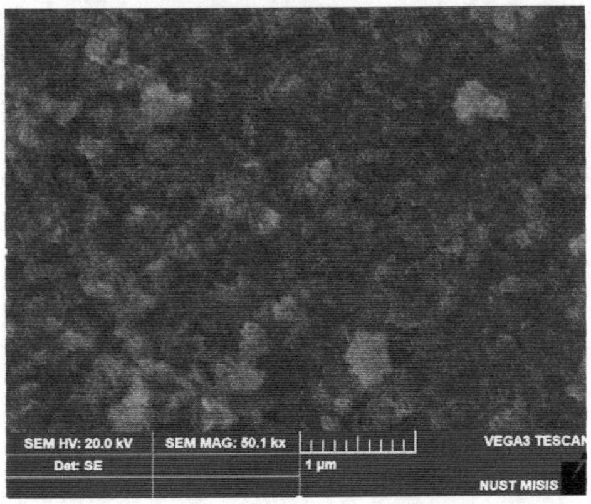

Figure 5. Microphotography of Co_3O_4 a sample with five nanolayers.

On the energy-dispersive X-ray spectrum are observed are observed cobalt, oxygen and a significant number of other elements characteristic of glass, that confirming the elemental composition of cobalt oxide agglomerates on the substrate. Figure 6 shows a multi-layer map of energy-dispersion spectroscopy, which suggests that Co_3O_4 nanolayers obtained by slot-die printing cover the entire surface of the sample.

Figure 6. Map of energy dispersion X-ray spectroscopy.

4. Conclusion

The proposed easy and cost-effective method of printing from dispersion makes it possible to form a uniform and dense cobalt oxide layer on the surface of substrates, which can be used prospectively in obtaining a hole-transport layer in perovskite solar cells.

References

[1] Ning Z, Fu Y and Tian H 2010 *Energy and Environmental Science* **3** 1170–1181

[2] Jeon N J, Lee J, Noh J H, Nazeeruddin M K, Gratzel M and Seok S I 2013 *Journal of the American Chemical Society* **135** 19087–19090

[3] Lee J H, Noh Y W, Jin I S, Park S H and Jung J W 2019 *Journal of Power Sources* **412** 425–432

[4] Bashir A, Shukla S, Haur L J, Shukla S, Bruno A, Gupta D, Baikie T, Patidar R, Akhter Z, Priyadarshi A, Mathews N and Mhaisalkar S G 2017 *The Royal Society of Chemistry* **10** 2341–2350

[5] Shalan A E, Oshikiri T,Narra S, Elshnawany M M, Ueno K, Wu H-P, Shi X, Diau E W-G and Misawa H 2016 *ACS Appied Materials and Interfaces* **8** 33592–33600

Modeling the atomic structure in the vicinity of the spherical voids and calculation of void growth rate anisotropy in bcc iron and tungsten

A V Nazarov[1,2], A P Melnikov[1], A A Mikheev[3] and I V Ershova[1]

[1] National Research Nuclear University MEPhI (Moscow Engineering Physics Institute), 31, Kashirskoe shosse 115409, Moscow, Russia
[2] Institute for Theoretical and Experimental Physics named by A.I. Alikhanov of NRC "Kurchatov Institute", 25 Bolshaya Cheremushkinskaya str.,117218, Moscow, Russia
[3] The Kosygin State University of Russia, 33 Sadovnicheskaya str., 117997, Moscow, Russia

E-mail: avn46@mail.ru

Abstract. We study the influence of the atomic structure in the vicinity of voids on their growth rate anisotropy. In the first part, we model the atomic structure in the vicinity of nanovoids in α-Fe and W using the advanced Molecular Statics method. In the second part, we use the earlier obtained equations that taking the influence of elastic fields into account to calculate the shifting rate of the void surface elements, and to evaluate the components of the strain tensor we use the atomic structure modeling results from the first part. The calculations are performed for voids of several sizes at certain oversaturation's in a wide temperature range. The simulation results for the mentioned metals with a bcc structure show that displacements of atoms located along the crystallographic directions of the <100>, <110>, <111> types in the vicinity of the voids are significantly different, and this anisotropy of atom displacements leads to a reduction of spherical symmetry for the shifting rate of the surface elements. As the result, the initially spherical void shape becomes faceted.

1. Introduction

During the irradiation of materials defects such as voids and bubbles are formed, and this changes the properties and characteristics of the material. Therefore, the study of the void growth kinetics is necessary to predict material behavior under work conditions, and these issues are under special attention in the works of many authors [1-5]. It is often seen at electron microscopy images that void, initially having the spherical shape, gradually change it to faceted [3, 4, 6]. According to the works [3, 7], several factors influence the change in the void shape upon irradiation: surface energy anisotropy for different crystallographic planes, preferred adsorption of atoms on certain crystallographic surfaces and vacancy diffusion fluxes anisotropy in the vicinity voids.

Ordinarily the elastic fields around the voids are not taken into account in models that predicting the material's behavior under irradiation [1]. However, as shown previously [8,9], these fields in the case of nanovoids have a significant effect on vacancy fluxes, and, consequently, on the growth and dissolution rates of voids. According to the results of [10] the X-axis component of the vacancy flux in the zero approximation of the elastic field effect is described by the following equation:

$$J_x = -D_V\left[\frac{\partial c}{\partial x} - c\frac{K^V}{kT}\frac{\partial Sp\varepsilon}{\partial x}\right], \tag{1}$$

where J_x is the vacancy diffusion flux in the direction of X-axis, c is the vacancy concentration, D_V is the vacancy diffusion coefficient in an ideal system, K^V is the coefficient that determines the elastic field contribution to the vacancy flux [10], ε is the strain tensor, $Sp\varepsilon$ is the strain tensor trace, k is the Botzmann constant, T is the temperature.

Usually displacement fields in the vicinity of defects are determined by solving the isotropic theory of elasticity equations. The solution of the equations for an isolated void of a spherical shape with radius R has the form [11]:

$$u_i = C_1\frac{x_i}{r^3}, \tag{2}$$

where u_i is the atom displacement from the initial position in the x_i direction, r is the distance from the center of the void, and C_1 determined by both the characteristics of the material and the void radius:

$$C_1 = -\frac{1+\nu}{2E}\gamma R^2, \tag{3}$$

where R is the radius of the spherical void, γ is surface energy, ν is the Poisson's ratio, E is the Young's modulus.

The components of the strain tensor are determined by the equation:

$$\varepsilon_{ij} = \frac{1}{2}\left(\frac{\partial u_i}{\partial x_j} + \frac{\partial u_j}{\partial x_i}\right). \tag{4}$$

When differentiate the displacement (Equation (2)) we get the equation for the diagonal components of the strain tensor:

$$\varepsilon_{ii} = C_1\left(\frac{1}{r^3} - \frac{3x_i^2}{r^5}\right), \tag{5}$$

and the trace of the strain tensor is a constant, and its derivative is zero:

$$\nabla Sp\varepsilon = 0. \tag{6}$$

Thus, elastic fields do not affect diffusion fluxes and void growth rates, and in that connection voids are often called neutral sinks [1]. In our work, we use the Advanced Molecular Static method for simulation atomic structure in the vicinity of the voids and finding atom displacements [12-15]. This approach takes into account the discreteness of the structure and its anisotropy. As a result, the atoms' displacements for different crystallographic directions differ significantly, and the trace of the strain tensor is different from the constant and its gradient is not zero. Therefore, an additional term appears in the equations for vacancy fluxes, and this term differ for various crystallographic directions.

The model is described in detail in [9], there are also some results for α-Fe. The results of modeling the structure are used to calculate in the next section the dependences of the shifting rate of the void surface elements on the oversaturation in metals with a bcc structure (α-Fe and W) over a wide temperature range. Moreover, these dependencies for crystallographic directions of types <100>, <110>, <111> are significantly different, and thus, the elastic field anisotropy in the vicinity of the void can lead to a change in the shape of the initially spherical voids in the irradiated materials.

2. Method

IOP Conf. Series: Materials Science and Engineering 1005 (2020) 012026 doi:10.1088/1757-899X/1005/1/012026

Atoms surround a spherical void with radius R, the positions of atoms are determined by the radius vector $\boldsymbol{r}$. The system is divided into two zones – main computational cell (zone I) and elastic medium in which atoms are immersed (zone II). The zone structure is shown on Figure 1.

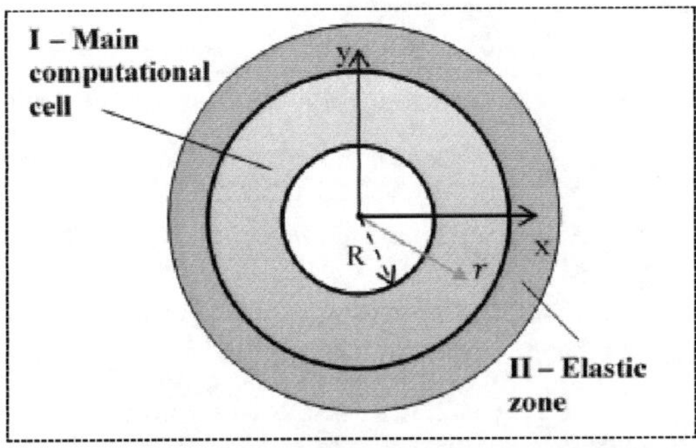

Figure 1. The location of the main computational cell and the elastic zone in the vicinity of a pore of radius R (cross section for z = 0).

The atomic coordinates of the first zone are calculated using the variational procedure of the MS method. The atoms surrounding the main computational cell are located in an elastic medium, and their displacements are determined based on solutions of the elasticity theory equations:

$$u=C_1\frac{\boldsymbol{r}}{r^3}. \tag{7}$$

An important feature of the model is a self-consistent iterative procedure for calculating the atoms' positions in the main computational cell and calculating the constant C_1, which determines the displacements in the elastic zone. The constant C_1 is calculated using equation (7) based on the results of modeling atomic displacements in the spherical layer located approximately in the middle between the defect and the boundary of the main computational cell. The simulation results indicate a stable convergence of the iterative procedure. The scheme of the iterative algorithm is presented on Figure 2.

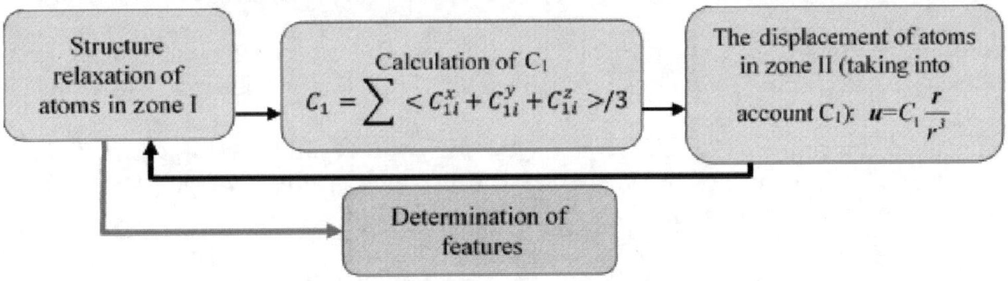

Figure 2. Iterative procedure scheme.

The model allows one to obtain a structure in the vicinity of voids [9,16]. The next section presents the simulation results for voids of various sizes in α-Fe and W that are further used in calculating the shifting rate of void surface elements.

3. Simulation results of atomic structure in vicinity of nanovoids

Figures 3 and 4 show the dependences of atomic displacements for crystallographic directions of the type <100>, <110>, <111> on the distance to the void center in α-Fe and W. Calculations of atomic displacements based on the solution of the isotropic theory of elasticity equation are also given for comparison for voids of the same size. These solutions have spherical symmetry and are independent of angles.

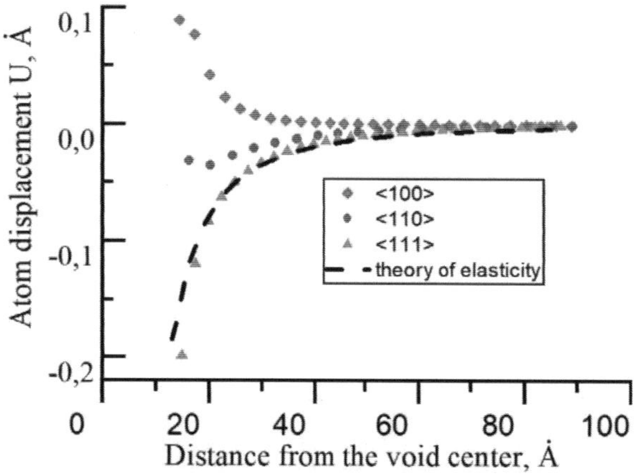

Figure 3. The atoms' displacements for various directions in α-Fe for nanovoid (R = 15.14 Å).

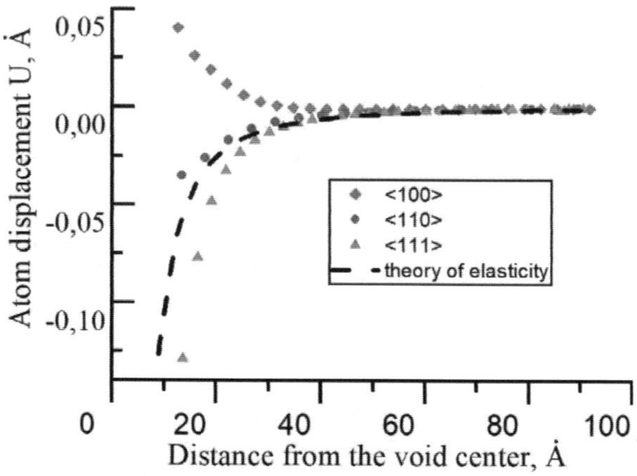

Figure 4. The atoms' displacements for various directions in W for nanovoid (R = 10.96 Å).

In contrast to the predictions of the isotropic theory of elasticity, atoms' displacements in different crystallographic directions differ significantly, both for α-Fe and W. For directions of type <100>, the displacements near the voids are positive. In α-Fe, as shown in Figure 3, a change in the sign of the displacement to negative occurs for atoms located from the void center by a distance of more than 42 Å (more than 12 lattice parameters). The maximum atoms' displacements from the site of the ideal lattice are observed in the direction <111>. It should be noted that according to the simulation results for voids of different sizes (up to 20 Å [16]), the dependences have a similar form.

From the analysis of the results it follows that the trace of the strain tensor is not zero, and the equations for vacancy fluxes, and consequently the kinetic equations for the void growth rate must contain additional terms due to the elastic field:

$$\nabla Sp\varepsilon \neq 0. \tag{8}$$

Therefore, in the next section, we present the derivation of equations for the shifting rate of void surface elements based on the expressions for the vacancy fluxes that we obtained earlier in [10].

4. Shifting rate of the void surface elements for the different crystallographic directions

It can be shown [17] that the movement of the shifting rate of void surface element in a certain direction is determined by the equation:

$$\frac{dR}{dt} = -\left(\vec{\boldsymbol{n}}, \vec{\boldsymbol{j}}\right), \tag{9}$$

where R is the radius of the void, $\boldsymbol{n}$ is the normal to the void surface, $\boldsymbol{j}$ is vacancy flux density to the void surface.

To evaluate the influence of the elastic field on the vacancy flux in the region near the voids and obtain analytical solutions, the method of successive approximations is used. And as a first approximation, similar to [18], we choose the solution of the diffusion equation for the vacancy concentration in that the influence of the field is not taken into account. Then the change rate of the void radius R is described by equation [18]:

$$\frac{dR}{dt} = C_{eq} D_V \left[\frac{R_G}{R_G - R} R^{-1} \left(\Delta + 1 - \exp\left(\frac{2\gamma V^f}{kTR} \right) \right) \right], \tag{10}$$

where c_{eq} is the equilibrium vacancy concentration for the flat surface, V^f is the vacancy formation volume, $\Delta = \frac{c_m - c_{eq}}{c_{eq}}$ is the oversaturation of vacancies, $c_m = c(R_G)$, where R_G is the average distance between voids.

Transformations for getting the equations for the shifting rate of the void surface elements for various crystallographic directions taking into account the field of elastic strains are given in [9]. Here, we restrict ourselves to providing final equations for the shifting rate of the void surface elements:

$$\frac{dR_{100}}{dt} = C_{eq} D_V \left[\frac{R_G}{R_G - R} R^{-1} \left(\Delta + 1 - \exp\left(\frac{2\gamma V^f}{kTR} \right) \right) - \frac{K^V}{kT} \exp\left(\frac{2\gamma V^f}{kTR} \right) \frac{\partial Sp\varepsilon}{\partial x} \right], \tag{11}$$

$$\frac{dR_{110}}{dt} = \frac{C_{eq} D_V}{\sqrt{2}} \left[\frac{\sqrt{2} R_G}{R_G - R} R^{-1} \left(\Delta + 1 - \exp\left(\frac{2\gamma V^f}{kTR} \right) \right) - \frac{K^V}{kT} \exp\left(\frac{2\gamma V^f}{kTR} \right) \left(\frac{\partial Sp\varepsilon}{\partial x} + \frac{\partial Sp\varepsilon}{\partial y} \right) \right], \tag{12}$$

$$\frac{dR_{111}}{dt} = \frac{C_{eq} D_V}{\sqrt{3}} \left[\frac{\sqrt{3} R_G}{R_G - R} R^{-1} \left(\Delta + 1 - \exp\left(\frac{2\gamma V^f}{kTR} \right) \right) - \frac{K^V}{kT} \exp\left(\frac{2\gamma V^f}{kTR} \right) \left(\frac{\partial Sp\varepsilon}{\partial x} + \frac{\partial Sp\varepsilon}{\partial y} + \frac{\partial Sp\varepsilon}{\partial z} \right) \right]. \tag{13}$$

The components of the strain tensor for each of the crystallographic directions, the trace of the strain tensor, and its derivatives to the corresponding coordinates are calculated based on the approximated simulation results of atom displacements near the void surface. After that we can calculate the shifting rate of the void surface elements using equations (11-13). It was done for various oversaturations in a wide temperature range. Figures 5, 6 show the normalized dependences of the mentioned void growth rates for α-Fe and W at the oversaturation $\Delta = 20$.

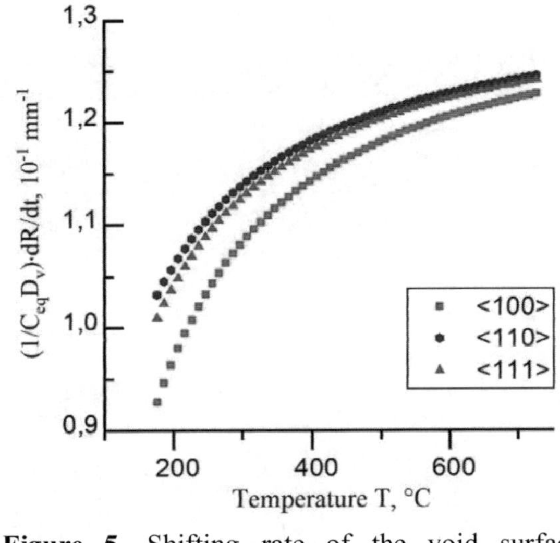

Figure 5. Shifting rate of the void surface elements for various crystallographic directions in α-Fe (R = 15.14 Å).

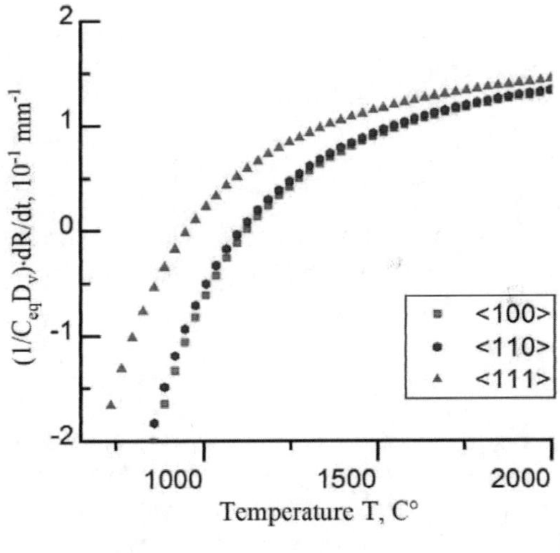

Figure 6. Shifting rate of the void surface elements for various crystallographic directions in W (R = 10.96 Å).

In α-Fe void growth rate are positive for all crystallographic directions, however, the void growth rate significantly slows down in the direction of the <100> type due to the negative contribution of the elastic field. For directions of the <110> and <111> types the growth rates are quite close. In W for the <111> direction the elastic displacement field makes a significant contribution to the shifting rate of the void surface element.

5. Discussion

As can be seen from equations (11-13), the additional term due to the elastic displacement field can contributes to the equations for the shifting rate of the void surface elements. According to the simulation results atomic displacements for α-Fe and W in the <100> direction are positive, and the corresponding contribution of the elastic field to shifting rate in equation (11) is negative. For this direction the void growth rate is slowed down, while the contributions in the directions <110> and <111> (see equations (12) and (13)) are positive.

In α-Fe, the shifting rate in the direction of the <100> type differs significantly from the directions of the <110>, <111> type. In W the atoms' displacements in the <100> direction also are positive, however displacements at void surface in W are notably less than in Fe, and then influence of elastic field to shifting rate of void surface element is lower.

Thus, for α-Fe and W, one of the reasons that influence the change in the shape of the initially spherical voids is the vacancy flux anisotropy due to the features of the atomic structure in the vicinity of the void. The asymmetry of atomic displacements in the vicinity of voids in α Fe and W slows down the shifting rate of the void surface elements in directions that coincide or are close to the crystallographic type <100>. The results show that initially the spherically symmetric void could change its shape, and due to the slowing down of the shifting rate of the void surface elements along the <100> directions, the void shape will gradually be faceted by {100} planes.

6. Conclusion

The effect of the anisotropy of atoms' displacements in the vicinity of nanovoids in α-Fe and W on the shifting rate of the void surface elements is studied. Using the Advanced Molecular Statics method, the atomic structure is simulated in the vicinity of voids in α-Fe and W. In contrast to the predictions of the isotropic theory of elasticity, the atoms' displacements in different crystallographic directions differ significantly, both for α-Fe and W. Should be emphasized that for directions of the type <100> displacements near the void are positive. Based on the simulation results for atoms' displacements, the dependences of the shifting rate of the void surface elements are calculated for various oversaturation's in a wide temperature range. Asymmetry of atoms' displacements in the vicinity of voids leads to a slowdown in the movement of surface elements in directions that are close to the <100>, that leads to a change in the void of the initially spherical shape to a shape faceted by {100} planes.

Acknowledgements

This work was supported by Competitiveness Growth Program of the Federal Autonomous Educational Institution of Higher Education National Research Nuclear University MEPhI (Moscow Engineering Physics Institute).

References

[1] Was G S *Fundamentals of radiation materials science* (New York: Springer) p 827.
[2] Garner F 2012 *Compr. Nucl. Mater.* **4** 33-95
[3] Chen C 1973 *Phys. Status Solidi.* **16** 197-210
[4] Zinkle S 2012 *Compr. Nucl. Mater.* **1** 65-98.
[5] Osetsky Y and Bacon D 2010 *Philos. Mag.* **90** 945-61
[6] Niwase K, Ezawa T, Fujita F, Kusanagi H and Takaku H 1988 *Radiat. Eff.* **106** 65-76
[7] Han G, Wang H, Lin D, Zhu X, Hu S and Song H 2017 *Comput. Mater. Sci.* **133** 22-34
[8] Mikheev A, Nazarov A, Ershova I and Zaluzhnyi A 2015 *Defect Diffus. Forum* **363** 91-98
[9] Nazarov A, Mikheev A and Melnikov A 2020 *Journal of Nuclear Materials* **532**, 152067–7
[10] Nazarov A and Mikheev A 2015 *Defect Diffus. Forum* **363** 112-119
[11] Eshelby J 1956 *Sol. State Phys.* **3** 79–144
[12] Valikova I, Nazarov A and Mikheev A 2006 *Defect Diffus. Forum* **249** 55-60
[13] Valikova I and Nazarov A 2008 *Defect Diffus. Forum* **277** 125-32
[14] Valikova I and Nazarov A 2008 *Phys. Met. Metallogr.* **105** 544-52
[15] Valikova I and Nazarov A 2010 *Phys. Met. Metallogr.* **109** 220-26
[16] Nazarov A, Ershova I and Volodin Y 2018 *KnE Mater. Sci.* **2018** 451-57
[17] Nazarov A Phys. 1973 *Met. Metallogr.* **35** (3) 184–186
[18] Cheremskoy P G, Slyozov V V and Betehtin V I 1990 *Voids in Solid State* (Moscow: Energoatomizdat) p 376 (in Russian)

Molecular static simulation of edge dislocation core in bcc iron

A A Gusev[1,2] **and A V Nazarov**[1,2]

[1] Institute for Theoretical and Experimental Physics named by A.I. Alikhanov of NRC "Kurchatov Institute", 25 Bolshaya Cheremushkinskaya str., 117218, Moscow, Russia
[2] National Research Nuclear University MEPhI (Moscow Engineering Physics Institute) 31 Kashirskoe shosse, 115409, Moscow, Russia

E-mail: avn46@mail.ru

Abstract. We simulate the dislocation core structure in bcc iron using the modified Molecular Static method. A feature of this method is the application of an iterative procedure in which the atomic structure in the vicinity of the defect and the constants that determine the displacements of atoms immersed in the elastic continuum are calculated in a self-consistent manner. Following the mentioned approach, we develop a model for calculating the atomic structure of edge dislocations, taking into account the anisotropy of the elastic medium surrounding the main calculation cell. Anisotropy is taken into account by introducing an explicit angular dependence for the parameters of the elastic field created by the dislocation: magnitude of Burgers vector and Poisson's ratio. Simulation is carried out for a split dislocation with Burgers vector along [100]. The convergence of the iterative algorithm is shown and the influence of the computational cell size on the results is considered. Calculated results are: atomic structure of dislocation in bcc iron, angular dependence of the parameters describing the elastic dislocation field at large distances from the dislocation line, and the strain tensor components in the entire simulation area.

1. Introduction

Segregation of certain elements in the vicinity of the dislocation lines alters the structure of the defect core that affects the ductility of the material. Therefore, the study of the distribution of elements near linear defects is of interest. A necessary prerequisite for this is information on the atomic structure of the dislocation core. In this regard, we simulate the structure of the core of the edge dislocation, and its surroundings using the modified method of molecular statics, taking into account the anisotropy of the elastic field at the boundary of the computational cell. The developed model makes it possible to match the displacements of atoms at the boundary of the main computational cell with the displacements of atoms located in an elastic medium, and calculate the components of the strain tensor in the entire computational domain. The obtained strains are supposed to be used in the future for modeling the kinetics of the redistribution of impurity atoms in the dislocation field.

2. Model description

As a modeling method, the method of Molecular Statics is used [1], as in a number of other works [2-4]. In this case, the main computational cell in which the atoms relax is surrounded by an elastic medium (zone), the atoms in which are displaced relative to the nodes of the ideal lattice in accordance with the solutions of the equations of elasticity theory for such a defect as a dislocation [2]. As a rule, in the aforementioned works it is not possible to ensure the correspondence of atomic displacements at the boundary of two zones — the main computational cell and the elastic zone. As a result of calculations, the difference in atomic displacements at the boundary of two zones can distort both the structure of the nucleus and the entire strain field in the vicinity of the dislocation. Therefore, the applied model has significant differences from the previously used ones and is a development of the model that was previously used to model point defects [5–8] and nanopores [9,10]. A common feature of these models is the application of an iterative self-consistent procedure for calculating the coordinates of atoms in the main cell and the parameters that determine the solutions of the equations of elasticity theory, the displacement of atoms immersed in an elastic medium. Moreover, the model used solutions for the case of dislocations in the approximation of an isotropic medium [2], and in the plane of the perpendicular line of the dislocation we have:

$$u_x = \frac{b}{2\pi}\left(\operatorname{atan}\left(\frac{y}{x}\right) + \frac{\pi}{2} sign(x) + \frac{xy}{2(1-\mu)(x^2+y^2)}\right), \quad (1)$$

$$u_y = -\frac{b}{4\pi}\left(\frac{1-2\mu}{2-2\mu}\ln(x^2+y^2) - \frac{y^2}{(1-\mu)((x^2+y^2)}\right), \quad (2)$$

where b – magnitude of Burgers vector, μ - Poisson ratio.

However, the crystal anisotropy is taken into account by determining the angular dependence of the Poisson coefficient and the Burgers vector determined during the iterations. Thus, the discrete nature of the crystal and the anisotropy of its properties are taken into account in the model. We add that the dislocation line is directed along the Z axis, and periodic boundary conditions are applied along this coordinate. System includes several atomic layers which are normal to Z axis. Burgers vector is oriented along X axis, and Y axis is oriented along line of extraplane's atoms. Simulation for every atomic layer is performed in similar way. The scheme for computational cell (one layer) and the main steps of the algorithm are shown in figure 1.

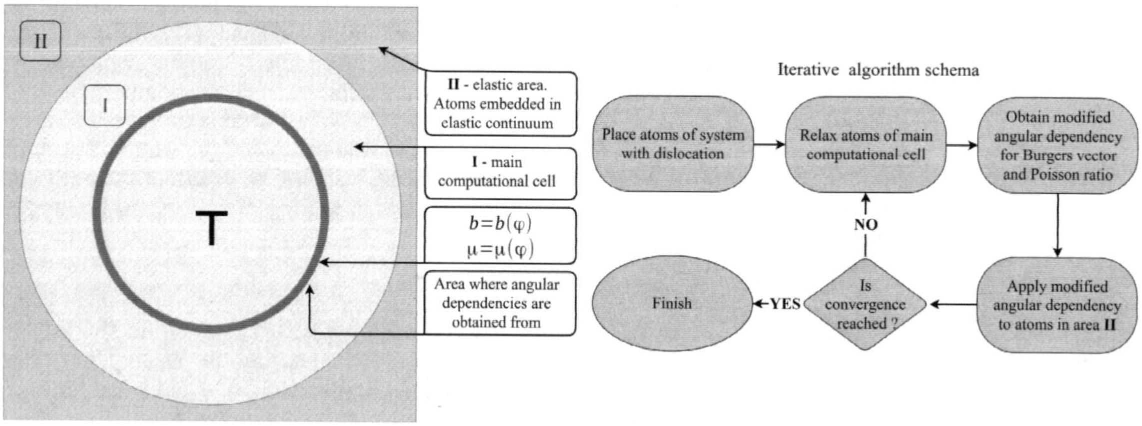

Figure1. The scheme of computational cell and the iterative algorithm steps.

The equilibrium positions of the atoms of the main computational cell are obtained using the variational procedure, like in usual molecular statics method [1]. An important feature of the algorithm that distinguishes it from usual versions of the Molecular Static Method is the application of the iterative procedure (see figure 1), in which the atomic coordinates of the main computational cell and two parameters are calculated in a self-consistent manner at each iteration step. These parameters are the Burgers vector and the Poisson ratio. Moreover, these parameters and their angular dependence are found on the basis of atomic displacements in the annular layer (see figure 1) calculated at the next iteration step by solving the inverse system of equations 1a and 1b. In this work, the structure of a split dislocation in bcc iron with the direction of the Burgers vector [100] is simulated using the interatomic potential of the EAM type [11]. A schematic structure of the dislocation is shown in figure 2.

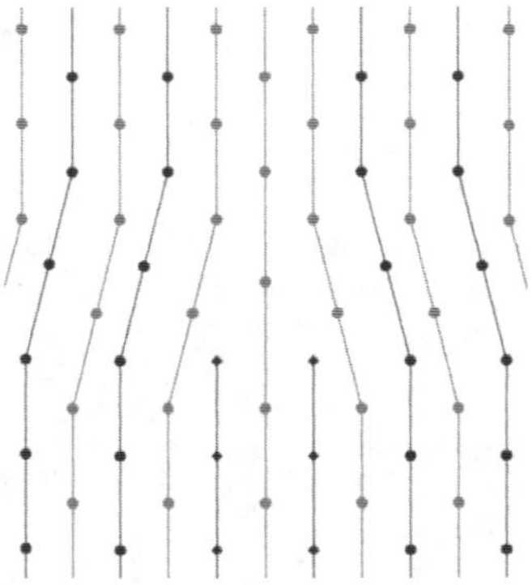

Figure 2. Scheme of a split dislocation structure.

3. Results and discussion

We test the model by studying the dependence of the results on the size of the system and the number of iterations. Simulation is performed for systems of several sizes (106 Å x 106 Å, 123 Å x 123 Å, 264 Å x 264 Å). Figure 3 and figure 4 show the angular dependences of the Burgers vector and the Poisson ratio on the number of iterations. Zero angle is set for Y axis (corresponds with extraplane).

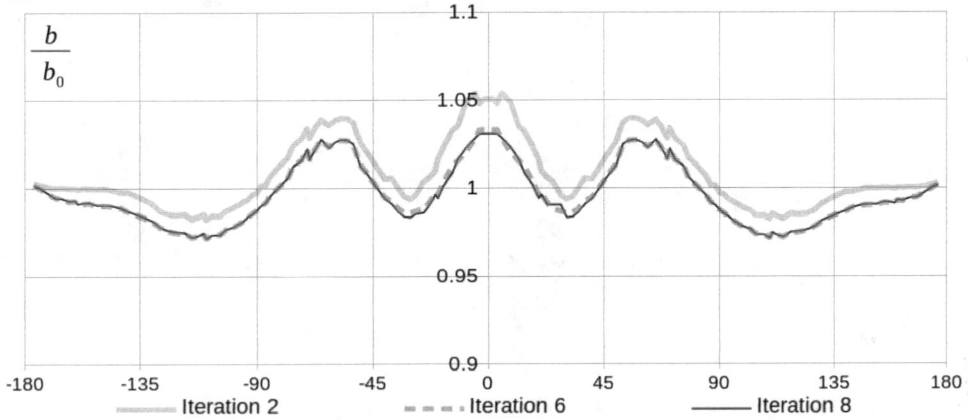

Figure 3. Normalized angular dependence of the Burgers vector magnitude for several iterations. b_0 =2.86 Å.

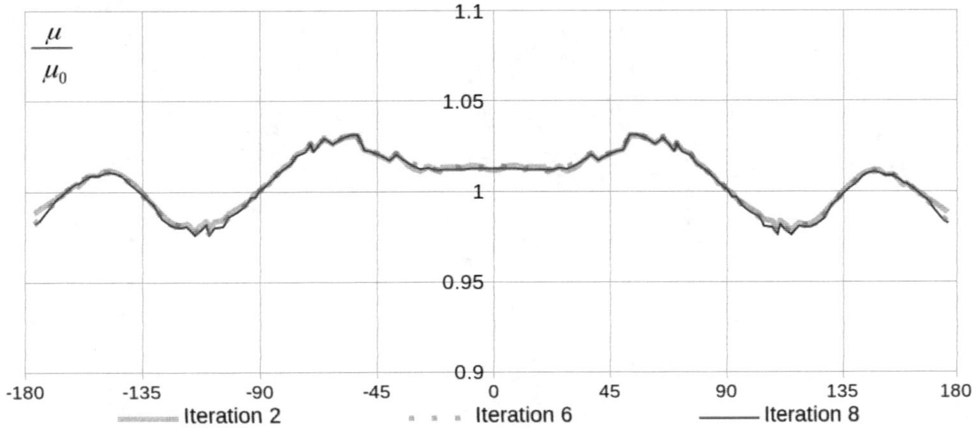

Figure 4. The normalized angular dependence of the Poisson's ratio for several iterations. μ_0 =0.369 [12].

IOP Conf. Series: Materials Science and Engineering 1005 (2020) 012027 doi:10.1088/1757-899X/1005/1/012027

Simulation shows that if size of the system is bigger than 132Å x 132 Å, then results are weakly dependent on size. The application of the developed model made it possible to calculate atomic displacements both in the main computational cell and in the elastic zone. It is worthy to emphasize that our approach grants good agreement of displacements at the zone boundary. Spatial distributions of the strain tensor components are calculated on the basis of displacements obtained during simulation, according to the known formulas:

$$\varepsilon_{ij} = \frac{1}{2}\left(\frac{\partial u_i}{\partial x_j} + \frac{\partial u_j}{\partial x_i}\right), \tag{3}$$

where indices *i* and *j*, take values of 1 and 2 due to symmetry. The corresponding dependence $\varepsilon_{11}(x,y)$ and $\varepsilon_{22}(x,y)$ in a plane normal to a line dislocation are shown at figure 5 and figure 6.

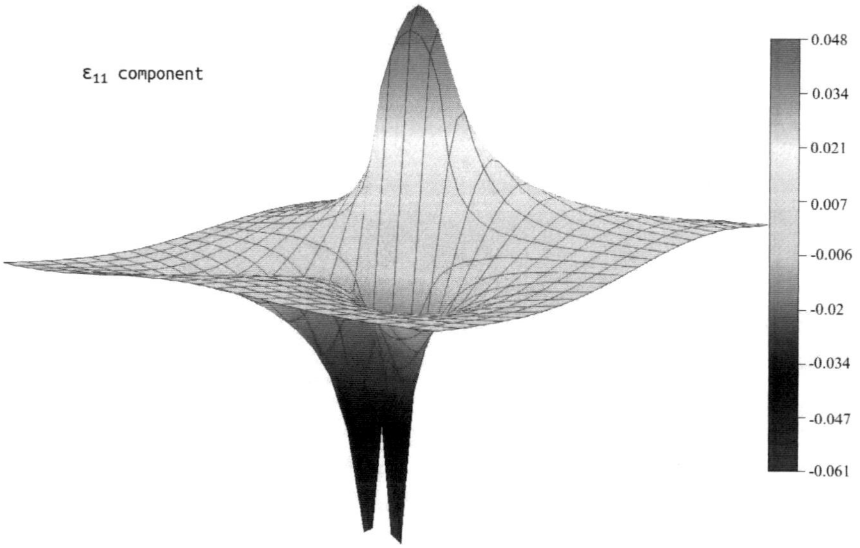

Figure 5. Spatial distribution of the tensor component $\varepsilon_{11}(x,y)$, *dl*=2.86.

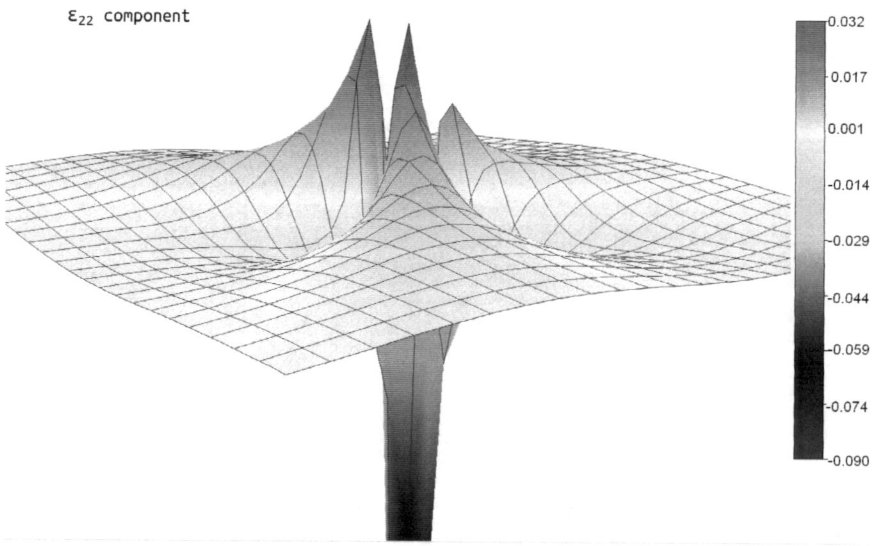

Figure 6. Spatial distribution of the tensor component ε_{22} *(x,y)*, *dl* =2.86.

Here *dl* is the step along the grid, the size of the shown area is 132 Å x 132 Å, and the calculations are performed for the area 264 Å x 264 Å. As it can be seen on figures, distributions have complex nonlinear character, especially near the dislocation line.

As it is indicated in [4], since the atoms embedded in elastic zone are usually fixed in their initial position and only the atoms of the main computational cell move, artifacts of elastic fields may take place at the boundary of these zones. The applied iterative algorithm modifies coordinates of atoms in zone II. This provides much better coordination of atomic displacements in the vicinity of the boundary between the main computational cell and the elastic zone. These features are very important in further solution of the equations that describe impurity atoms redistribution in the elastic field of dislocations, where diffusion coefficients are nonlinearly dependent on strain tensor components.

4. Conclusion

We developed model that is based on Modified Method of Molecular statics, that allows to calculate atomic structure of dislocation core and its vicinity. The feature of model is self-consistent iterative procedure for determining the coordinates of atoms in main computational cell and parameters that define displacements of atoms in elastic zone. Moreover, model allows taking anisotropy of the crystal structure into account, because these parameters depend on the angle.

The atomic structure of a split dislocation with a Burgers vector [100] in bcc iron is simulated, and strain tensor components are calculated.

The applied iterative algorithm gives good compatibility of atomic displacements in the vicinity of boundary between main computational cell and elastic zone.

Results of simulation show that distributions of strain tensor components in the vicinity of the dislocation line have a complicated nonmonotonic character.

Acknowledgements

This work was supported by Competitiveness Growth Program of the Federal Autonomous Educational Institution of Higher Education National Research Nuclear University MEPhI (Moscow Engineering Physics Institute).

References

[1] Johnson R A 1964 *Phys. Rev.* **134** (5A) 1329–36
[2] Bulatov V V and Cai W 2006 *Computer Simulations of Dislocations* (Oxford Series on Materials Modelling)
[3] Osetsky Yu N and Bacon D J 2003 *Modelling Simul. Mater. Sci. Eng.* **11** 427– 46
[4] Hartley C S and Mishin Y 2005 *Acta Materialia* **53** 1313–21
[5] Valikova I V and Nazarov A V 2008 Defect Diffus. Forum **277** 125–130
[6] Valikova I V and Nazarov A V 2008 *Phys. Met. Metallogr.* **105** (6) 544–52
[7] Valikova I V and Nazarov A V 2010 *Phys. Met. Metallogr.* **109** (3) 220–26
[8] Valikova I V, Nazarov A V et al 2011 *Solid State Phenomena* **172–174** 1156–63
[9] Nazarov A V, Ershova I V and Volodin Y S 2018 *KnE Materials Science* 451–7
[10] Nazarov A V, Mikheev A A and Melnikov A P 2020 *Journal of Nuclear Materials* **532**, 152067-7
[11] Ackland G J, Mendelev M I, Srolovitz D J, Han S and Barashev A V 2004 *J .Phys. Condens Matter* **16** (27) 2629–42
[12] Ledbetter H M and Reed R P 1973 *J. Phys. Chem. Ref. Data* **2** (3) 531–617

IOP Conf. Series: Materials Science and Engineering 1005 (2020) 012028 doi:10.1088/1757-899X/1005/1/012028

Kinetics of segregation formation in elastic field of edge dislocation in bcc iron

A A Gusev [1,2] **and A V Nazarov**[1,2]

[1] Institute for Theoretical and Experimental Physics named by A.I. Alikhanov of NRC "Kurchatov Institute", Bolshaya Cheremushkinskaya 25, 117218, Moscow, Russia
[2] National Research Nuclear University MEPhI (Moscow Engineering Physics Institute), Kashirskoe highway 31, Moscow, 115409, Russia

E-mail: avn46@mail.ru

Abstract. We study the kinetics of the redistribution of impurity atoms in the elastic fields of dislocations. Study is based on taking into account the influence of elastic fields on diffusion fluxes of interstitial impurity atoms. The study is performed by computer simulation methods and consists of several stages. The first is the simulation of a dislocation core structure with a Burgers vector along [100] direction by using the modified molecular static method. The second is obtaining the coefficients that determine the influence of the components of the strain tensor on the diagonal elements of the matrix of diffusion coefficients and in the equations for fluxes of carbon atoms in bcc iron. The third stage is associated with modeling the diffusion characteristics of carbon atoms by the method of molecular dynamics in a crystal without defects. The fourth and final stage uses the data about atomic structure that we obtained at first stage, as well as the characteristics calculated in the second and third, is a simulation of the formation of segregations of interstitial atoms, which is based on solving diffusion equations that take into account the elastic deformations created by the dislocation. The complex of developed models is used to analyze the kinetics of segregation formation. 3D graphs illustrate the distribution of interstitial atoms in the vicinity of a dislocation for different times at certain temperatures.

1. Introduction

Many properties of materials depend on the behavior of dislocations, which is largely determined by the structure of these defects, as well as by segregation of atoms in the vicinity of these defects [1-3]. In this regard, simulation of elements redistribution kinetics near linear defects is of interest for properties materials estimation.

Our approach is based on a theory that takes into account the effect of elastic fields on diffusion fluxes [4-7] and the results of modeling of the atomic structure of a dislocation core and its surroundings by us [8]. We use multiscale model, which includes several stages of simulation.

The first stage that was already implemented in [8] is the study of the atomic structure of the core of the edge dislocation using the modified molecular statics method that takes into account the anisotropy of the elastic field in the medium surrounding the computational cell.

The second stage is devoted to calculating the *Strain Influence on Diffusion (SID)* coefficients [5,6] that determine the influence of the components of the strain tensor on the fluxes of carbon atoms in the bcc structure, and the diffusion characteristics of these atoms are calculated. SID coefficients are extremely sensitive to the atomic structure in the vicinity of the defect, and especially to the atomic structure in the saddle position. Therefore, for their calculation, we use the modified method of Molecular Statics (MMoMS), which we developed earlier in [9–15]. The mentioned coefficients and diffusion characteristics are calculated for carbon atoms in bcc iron using paired and EAM potentials [16-18].

At the third stage, we model the diffusion jumps of carbon atoms in bcc iron using the molecular dynamics method and calculate the migration energy and preexponential factor for the Arrhenius equation.

The final stage is a numerical solution of the diffusion equation that determines the kinetics of the formation of impurity segregation in the vicinity of the edge dislocation. The equation takes into account the nonlinear dependence of the diffusion coefficients on strain tensor components. The values obtained in the second and third stages, as well as the strain field obtained by us earlier [8], are used to determine spatial distribution of diffusion coefficients in the vicinity of the dislocation. The redistribution of carbon atoms in the vicinity of edge dislocations with Burgers vector along [100] is simulated.

The results show that the distribution of impurity atoms is complex and the approach proposed in this work gives us an ability to study the kinetics of segregation formation in the vicinity of dislocations.

2. Simulation of atomic structure of edge dislocation core and it's vicinity

We use Modified Method of Molecular Static (MMoMS). In this version of MS method, the main computational cell, where atoms relax, is surrounded by an elastic zone, where atoms are displaced from the ideal lattice sites in accordance with the solutions of the equations of elasticity theory. This approach, which was previously used to model point defects [9–13] and nanopores [14,15], is now adapted to find the structure of the dislocation core and its surroundings. A common feature of these models is the application of an iterative, self-consistent procedure for calculating the coordinates of atoms in the main cell and the parameters that determine the displacements of atoms immersed in an elastic medium. We believe that displacements are given in solutions of equations of the elasticity theory. Moreover, in the developed version of the model, solutions used for dislocation in the approximation of an isotropic medium, and the anisotropy is taken into account by determining the angular dependence of the Poisson coefficient and the Burgers vector determined during the iterations (now these are refined parameters).

Application of the developed model made it possible to calculate atomic displacements both in the main computational cell and in the elastic zone. A detailed description of the model and results is given in [8]. Based on the components of atomic displacement vectors obtained during simulation, the components of the strain tensor, which are used later, are calculated.

3. Equations for carbon flux density under strain and calculation of SID coefficients

The equations for the flux density of carbon atoms, taking into account the influence of the elastic field, were obtained by us earlier [6]:

$$J_1 = -\frac{1}{\Omega}\left(D_1\frac{\partial c}{\partial x} + c\frac{\partial D_1}{\partial x}\right), \qquad J_2 = -\frac{1}{\Omega}\left(D_2\frac{\partial c}{\partial y} + c\frac{\partial D_2}{\partial y}\right), \tag{1}$$

where c is an impurity concentrartion, Ω is volume per atom.

For the third axis, the equation is similar. The diagonal elements of the matrix of diffusion coefficients have the form:

$$D_1 = \frac{D}{2}\exp\left(-\frac{K_1\varepsilon_{11}}{kT}\right)\left[\exp\left(-\frac{K_2\varepsilon_{22}+K_3\varepsilon_{33}}{kT}\right) + \exp\left(-\frac{K_2\varepsilon_{33}+K_3\varepsilon_{22}}{kT}\right)\right], \tag{2}$$

where D is the diffusion coefficient of impurity in absence of any other defects, $\varepsilon_{ij} = \frac{1}{2}(\frac{\partial u_i}{\partial x_j} + \frac{\partial u_j}{\partial x_i})$ is the strain tensor (i, j = 1,2,3), u_i are displacement vector components, k is Boltzmann constant , T is absolute temperature.
D_2 has similar form, that one may obtain by cyclic change of indices 1,2,3.

It is easy to see that each of these coefficients depends nonlinearly on the components of the strain tensor, and this functional dependence is determined by the SID coefficients. For the case of diffusion of a carbon atom in a bcc metal (jumps of the O - T - O type), three coefficients K_1, K_2, K_3 determine the effect of strains on diffusion, and they are linear combinations of coefficients [6]:

$$K_{xx}^W = \frac{1}{2}\sum_s \sum_{k\neq s} \frac{(x_{ks}^W)^2}{R_{ks}^W} \frac{\partial E}{\partial R_{ks}}\bigg|_{R_{ks}^W}, \; K_{xx}^0 = \frac{1}{2}\sum_s \sum_{k\neq s} \frac{(x_{ks}^0)^2}{R_{ks}^0} \frac{\partial E}{\partial R_{ks}}\bigg|_{R_{ks}^0}, K_{xy}^W = \frac{1}{2}\sum_s \sum_{k\neq s} \frac{x_{ks}^W y_{ks}^W}{R_{ks}^W} \frac{\partial E}{\partial R_{ks}}\bigg|_{R_{ks}^W} \quad (3)$$

and analogous to them. There x_k, y_k, z_k are coordinates of k-th atom, x_s, y_s, z_s are coordinates of s-th atom, $x_{ks} = x_k - x_s$, $y_{ks} = y_k - y_s$, $z_{ks} = z_k - z_s$, $k \neq s$, $R_{ks} = |\boldsymbol{r}_k - \boldsymbol{r}_s| = \sqrt{x_{ks}^2 + y_{ks}^2 + z_{ks}^2}$ for all atoms in system, upper index W relates to system configuration at saddle position of jumping atom, upper index 0 relates to system in equilibrium state before jump occurs, E is the energy of system.

For SID coefficients in the system, we have [6]:

$$K_1 = K_{xx}^W - K_{xx}^0, \quad K_2 = K_{yy}^W - K_{yy}^0, \; K_3 = K_{yy}^W - K_{xx}^0. \quad (4)$$

At this stage, the coefficients are calculated that determine the influence of the components of the strain tensor on the flows of defects. These coefficients are very sensitive to the atomic structure of the vicinity of the defect, and especially to the positions of atoms in the saddle position. For this reason, we use the MMoMS developed by us earlier [9–13], in which the atomic structure in the vicinity of the defect and the parameter that determines the displacements of atoms in the elastic zone are calculated in a self-consistent manner. A detailed description of the model originally developed for the vacancy diffusion mechanism is given in the mentioned works. In the simulation, multiparticle [16] and pair potentials [17,18] are used. Atomic structures in the vicinity of the carbon atom located in the initial position before the jump and saddle point are found during modeling. The model made it possible to calculate not only the energy of migration of the carbon atom, but also the relaxation volumes V_R and migration volumes V_m, and also SID coefficients K_1, K_2, K_3.

The simulation results of this stage are given in the Table 1.

Table 1. The SID coefficients and others characteristics of the carbon atom

Potentials	K_1, eV	K_2, eV	K_3, eV	Q, eV	V_R / Ω	V_m/Ω
[17,18]	7.31	-3.43	3.33	0.86	0.377	0.09
[16,18]	5.67	-4.10	2.89	0.87	0.176	0.18

4. Molecular dynamics simulation of carbon atom diffusion

In carrying out this part of the work, the "natural thermostat" model [19] was used that is based on a combination of molecular dynamics (MD) methods and molecular statics, and takes into account the elastic displacements of metal atoms at large distances from the defect and the temperature dependence of these displacements. The concept of a natural thermostat implemented in the model allows one to avoid some negative aspects inherent in models with periodic boundary conditions when studying diffusion jumps of atoms. The model diagram is shown in figure 1.

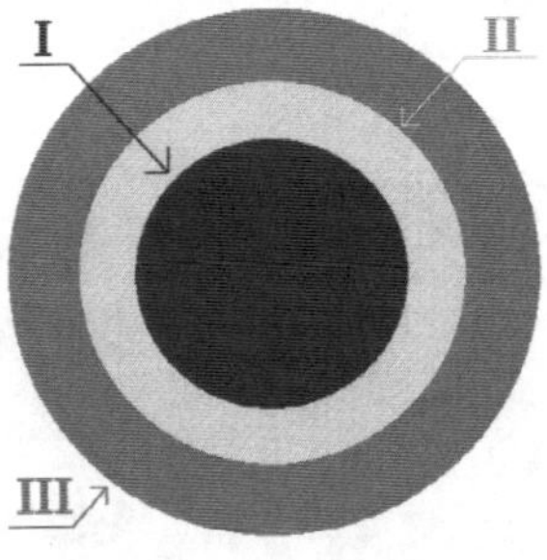

Figure 1. The scheme of the computational cell in MD modeling.
Zone I. Atoms move in accordance with the solutions of MD differential equations
Zone II. Thermostat area
Zone III. Atoms are motionless. Their coordinates are displaced from the ideal lattice sites in accordance with the solution of the isotropic elasticity theory.

Zone II is a spherical layer located around zone I. The equations of motion for zone II have the same form as for zone I. The difference is that the atomic velocity at each step is corrected using the coefficient β, which is calculated by the following formula:

$$\beta = [3NkT/\textstyle\sum_i mv_i^2]^{1/2} \tag{5}$$

as it is usually being done in MD simulations in the case of the isokinetic method (ensemble) [20]. The atoms of zone II interact in the usual way with both atoms of zone I and atoms of zone 3.

Zone III. Atoms are fixed. The atomic coordinates in zone III are set according to the structure of the crystal containing the defect in accordance with the spherically symmetric term of solutions from elasticity theory. At large distances from defect and taking into account thermal expansion (like in [21]):

$$u_x = C_1\frac{x}{r^3} \quad u_y = C_1\frac{y}{r^3}, \quad u_z = C_1\frac{z}{r^3}, \quad C_1 = C_{10}(1+\alpha T)^3, \tag{6}$$

where α is the coefficient of thermal expansion. The thermal expansion coefficients for various metals and the temperature dependence (6) were obtained in [21] for MD modeling, and C_{10} was calculated by the MMoMS in the previous section for T = 0.

Such a division into zones in this model is the closest to the conditions of a natural experiment, when the sample is in a thermostat. This gives ability to simulate various processes in crystals at constant temperature. Here, the role of the sample is played by zone I, and the role of the thermostat is played by zone II. This characteristic feature of our model reduces distortion of the fluctuation spectrum in zone I, in contrast to existing methods for modeling the canonical ensemble [20].

The calculation of the diffusion characteristics of carbon, based on the MD simulation of atomic jumps, is carried out according to the following algorithm. For each temperature, a set of time intervals that separate successive atomic jumps is obtained, the inverted intervals of these jumps are calculated, and then the average frequency Γ. This made it possible to determine the diffusion coefficient by the well-known formula [22]:

$$D = \frac{1}{6}l^2\Gamma, \tag{7}$$

where l is the length of the jump. Based on the simulation results, the temperature dependence of the diffusion coefficients is determined for two different interaction potentials of iron atoms [16,17].

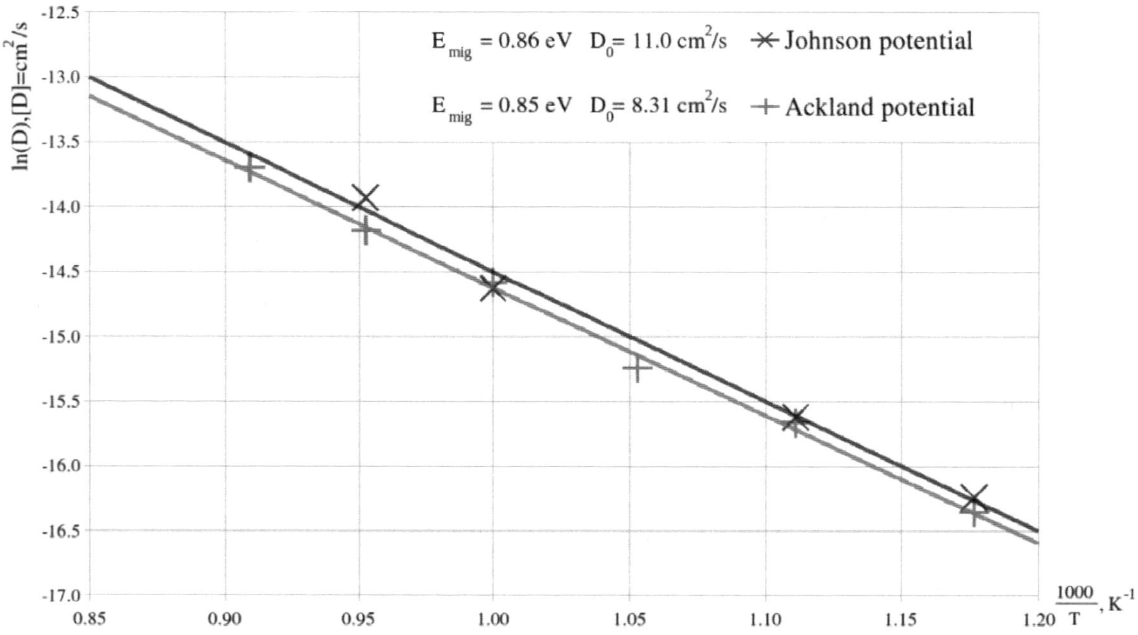

Figure 2. Arrhenius dependence of the diffusion coefficients of carbon in bcc iron.

Characteristics of diffusion coefficients and data from other works are given in Table 2.

Table 2. The migration energy of carbon atoms E_m and preexponential factor D_0, obtained by various modeling methods and experimental data.

Method / Potential	D_0, $10^{-7}m^2s^{-1}$	E_m, eV	Reference
MD / [16,18]:	8.31	0.85	This work
MS / [16,18]	-	0.87	This work
MD / [16,18]	1.89 ± 0.01	0.71 ± 0.02	[24]
MS / [16,18]	-	0.83	[24]
Experiment:	7.0	0.87	[25]

5. Simulation of segregation formation in vicinity of edge dislocation

In this section, we first obtain the diffusion equation that takes into account the effect of the elastic field on the flux. To do this, we need to substitute the expressions for flows into the continuity equation.

The redistribution of embedded carbon atoms is simulated by numerically solving the diffusion equation with allowance for the strains created by the dislocation in the elements of the matrix of diffusion coefficients (equation 3). As a result, we find the concentration $c(x,y,t)$ in the plane normal to the dislocation line. The initial concentration of interstitial atoms is assumed to be the same and equal to c_0. The calculations are performed for regions of different sizes and, starting from a region of size (261 Å x 261 Å), the results are almost indistinguishable. Examples of the distribution of the concentration of carbon atoms near the dislocation are shown in Figures 3 and 4.

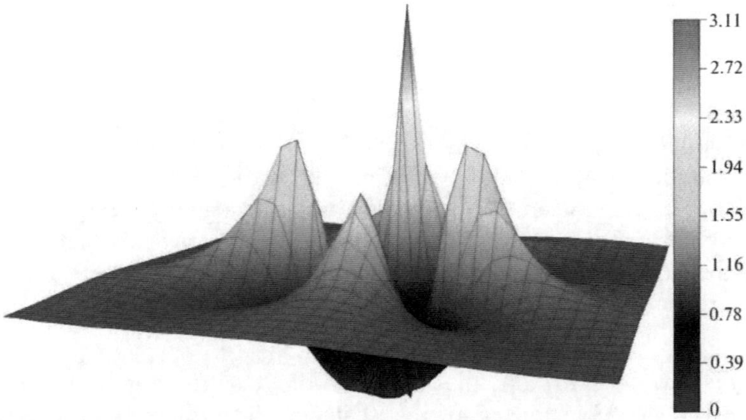

Figere 3. Normalized concentration of carbon in vicinity of edge dislocation, elapsed time t = 8 ns, dl = 2.86Å, $c^{\max}/c_0$ =3.11, T=973K, dl is the grid step, shown area is (143 Å x 143 Å).

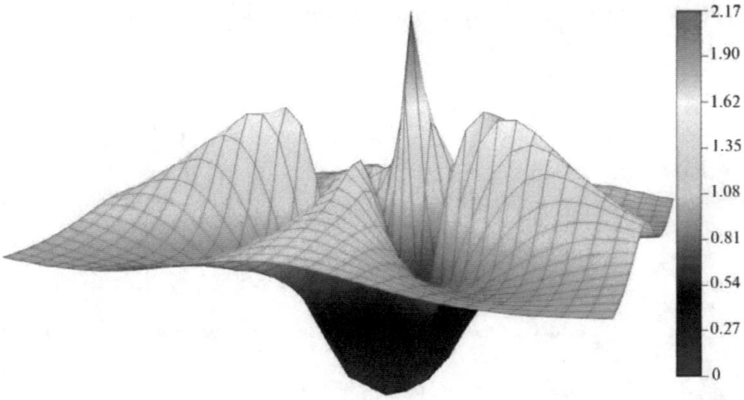

Figure 4. Normalized concentration of carbon in vicinity of edge dislocation, elapsed time t = 8 ns, dl = 2.86Å, $c^{\max}/c_0$ =2.17, T=1173K, dl is the grid step, shown area is (143 Å x 143 Å).

6. Discussion

The simulation results show that the distribution of carbon atoms near the dislocation line has a very complex irregular character. The nature of the distribution depends on the strain field created by the dislocation, on the SID coefficients, their relationship with the activation barrier in an ideal crystal, and also on temperature. Table 3 shows spatial maximums of carbon concentrations found for different time at a temperature of T = 1173 K, and in Table 4 at T = 973 K.

Table 3. The dependence of the maximum carbon concentration on time. T = 1173 K.

	t_1	t_2	t_3	t_4	t_5	t_6
t, ns	0.08	0.26	8	50	102	377
c^{max}/c^0	2.60	2.80	2.17	1.62	1.53	1.42
$10^6\ C_{min}/c_0$	145.7	90.8	42.4	36.5	31.8	29.8

Table 4. The dependence of the maximum carbon concentration on time. T = 973 K.

t, ns	0.08	0.26	8	50	102	377
c^{max}/c^0	2.67	2.77	3.11	2.40	2.07	1.71
$10^6\ C_{min}/c_0$	42.2	20.9	6.45	4.30	3.81	3.26

IOP Conf. Series: Materials Science and Engineering 1005 (2020) 012028 doi:10.1088/1757-899X/1005/1/012028

From the above data it follows that the concentration at the maximum increases rapidly at the beginning, and then decreases rather slowly. This feature of the formation of Cottrell atmospheres should probably be taken into account when assessing their influence on the dislocation movement.

Simulation for different initial concentrations shows that the ratio of concentrations at different coordinates to the initial one does not change. It is characteristically that the conformity in the distribution of concentrations is kept. This is why figures show relative (normalized) values.

The maximum concentration near the dislocation line noticeably exceeds the corresponding one obtained by modeling the kinetics of formation of carbon segregations in nickel [26]. In that work, for the components of the strain tensor, the results of the theory of elasticity outside the dislocation core were used, and the structure of the dislocation core was not simulated.

7. Conclusion

A multiscale model is developed to study the kinetics of the redistribution of impurity atoms in elastic fields of dislocations. MMoMS characteristics are calculated that determine the effect of the elastic field on the matrix of diffusion coefficients of carbon atoms in bcc iron, as well as the value of the activation barrier. Then, using the MD method, we simulated carbon migration at various temperatures and determined the migration energy and the preexponential factor for diffusion coefficients of carbon in bcc iron. In conclusion, the diffusion equation that determines the kinetics of the redistribution of carbon atoms in the vicinity of the dislocation for various temperatures is numerically solved. The simulation uses results of atomic displacements calculations in the core of dislocation and its vicinity, obtained in our work [8] and calculated therein, the components of the strain tensor.

The results indicate that the distribution of carbon concentration in the vicinity of the dislocation line has a complex nonmonotonic character and quickly reaches its maximum in the closest position to the dislocation line, then slowly decreases in this position, and the distribution minima and maxima in the vicinity of the line are irregular with complex geometry.

References

[1] Bulatov V V and Cai W 2006 *Computer Simulations of Dislocations (Oxford Series on Materials Modelling)*

[2] Blavette D, Cadel E, Fraczkiewicz A and Menand A. 1999 *Science* **286** 2317–9

[3] Wilde J, Cerezo A and Smith G D W 2000 *Scripta Materialia* **43** (1) 39–48

[4] Nazarov A V and Mikheev A A 2004 *Physica Scripta* **T108** 90–94

[5] Nazarov A V and Mikheev A A 2008 *J. Phys. Condens. Matter* **20** 485203.1

[6] Nazarov A V, Mikheev A A, Valikova I V and Zaluzhnyi A G 2011 *Solid State Phenomena* **72–174** 1156–1163

[7] Nazarov A V and Mikheev A A 2015 *Defect Diffusion Forum* **363** 112-9

[8] Gusev A A and Nazarov A V 2020 *this issue*

[9] Valikova I V, Nazarov A V and Mikheev A A 2006 *Defect Diffusion Forum* **249** 55–60

[10] Valikova I V and Nazarov A V 2008 *Defect Diffusion Forum* **277** 125–132

[11] Valikova I V and Nazarov A V 2008 *Phys. Met. Metallogr.* **105** (6) 544–52

[12] Valikova I Vand Nazarov A V 2010 *Phys. Met. Metallogr.* **109** (3) 220–6

[13] Valikova I V and Nazarov A V 2011 *Solid State Phenomena* 1222–7

[14] Nazarov A V, Ershova I V and Volodin Y S 2018 *KnE Materials Science* 451–457

[15] Nazarov A V, Mikheev A A and Melnikov A P 2020 *Journal of Nuclear Materials* **532**, 152067–7

[16] Ackland G J, Mendelev M I, Srolovitz D J, Han S and Barashev A V 2004 *J.Phys. Condens Matter* **16** (27) 2629–42

[17] Johnson R A 1964 *Phys. Rev.* **134** (5A) 1329–1336

[18] Johnson R A, Dienes G J and Damask A C 1964 *Acta Metallurgica* **12** 1215–24

[19] Prokopenko D 2013 *Diploma MEPhI*

[20] Lesar R 2013 *Introduction to Computational Materials Science: Fundamentals to Applications*

[21] Germanov A B, Ershova I V, Kislitskaya E V, Nazarov A V and Zaluzhnyi A G 2013 *Metallofizika I Noveishiye Technologii (Metal Physics and Advanced Technologies)* **35** (10) 1391–1403
[22] Mehrer H 2007 *Diffusion in solids – Fundamentals, methods, materials, diffusion-controlled processes* (Springer-Verlag, Berlin Heldeberg)
[23] Gusev A A 2018 *Diploma MEPhI*
[24] Tapasa K, Barashev A V, Bacon D J and Osetsky Yu N 2007 *Acta Materialia* **55** 1–11
[25] Haneczok G, Weller M and Diehl J 1993 *Materials Science Forum* **119-121** 101–7
[26] Mikheev A A, Nazarov A V, Ershova I V and Zaluzhnyi A G 2016 *IOP Conference Series: Materials Science and Engineering* **130** (1) 01—062.

AUTHOR INDEX

Institute of Physics
Dirac House, Temple Back
Bristol BS1 6BE UK

ISSN: 1757-8981
ISBN 978-1-7138-3317-8
9 781713 833178